AF864382

Springer Proceedings in Mathematics & Statistics

Volume 430

This book series features volumes composed of selected contributions from workshops and conferences in all areas of current research in mathematics and statistics, including data science, operations research and optimization. In addition to an overall evaluation of the interest, scientific quality, and timeliness of each proposal at the hands of the publisher, individual contributions are all refereed to the high quality standards of leading journals in the field. Thus, this series provides the research community with well-edited, authoritative reports on developments in the most exciting areas of mathematical and statistical research today.

Christos P. Kitsos • Teresa A. Oliveira •
Francesca Pierri • Marialuisa Restaino
Editors

Statistical Modelling and Risk Analysis

Selected contributions from ICRA9, Perugia, Italy, May 25-27, 2022

Springer

Editors
Christos P. Kitsos
Univeristy of West Attica
Egaleo, Greece

Teresa A. Oliveira
Universidade Aberta
Lisboa, Portugal

Francesca Pierri
University of Perugia
Perugia, Italy

Marialuisa Restaino
University of Salerno
Fisciano, Italy

ISSN 2194-1009 ISSN 2194-1017 (electronic)
Springer Proceedings in Mathematics & Statistics
ISBN 978-3-031-39863-6 ISBN 978-3-031-39864-3 (eBook)
https://doi.org/10.1007/978-3-031-39864-3

Mathematics Subject Classification: 62-06, 62P10, 62P12, 62P20

© The Editor(s) (if applicable) and The Author(s), under exclusive license to Springer Nature Switzerland AG 2023
This work is subject to copyright. All rights are solely and exclusively licensed by the Publisher, whether the whole or part of the material is concerned, specifically the rights of translation, reprinting, reuse of illustrations, recitation, broadcasting, reproduction on microfilms or in any other physical way, and transmission or information storage and retrieval, electronic adaptation, computer software, or by similar or dissimilar methodology now known or hereafter developed.
The use of general descriptive names, registered names, trademarks, service marks, etc. in this publication does not imply, even in the absence of a specific statement, that such names are exempt from the relevant protective laws and regulations and therefore free for general use.
The publisher, the authors, and the editors are safe to assume that the advice and information in this book are believed to be true and accurate at the date of publication. Neither the publisher nor the authors or the editors give a warranty, expressed or implied, with respect to the material contained herein or for any errors or omissions that may have been made. The publisher remains neutral with regard to jurisdictional claims in published maps and institutional affiliations.

This Springer imprint is published by the registered company Springer Nature Switzerland AG
The registered company address is: Gewerbestrasse 11, 6330 Cham, Switzerland

Paper in this product is recyclable.

Preface

It is the ICRA's believe, and we repeat it also in this volume: Everything humans venture to do has some degree of risk involved in it. Even in our everyday life. That might be one of the reasons we are devoted to studying Risk, despite the risk. The plethora of research papers in the field is getting lager, and this might be one of the explanations that all the ICRA conferences are successful.

In principle Risk is defined as an exposure to the chance of injury or loss. Practically, it is a hazard or dangerous chance and is wondering about the probability that something unpleasant will take place. Therefore, the probability of damage, caused by external or internal factors has to be evaluated. The essential factors that influence the increment of the Risk are asked to be determined. That is why eventually we are referring to Relative Risk (RR).

Under this line of thought, we started the ICCRA (= International Conference on Cancer Risk Assessment) conferences on August 22, 2003, in Athens and we proceeded in Santorini, 2007 and 2009. We moved to Limassol, Cyprus 2011, with the essential adjustment to ICRA (= International Conference to Risk Analysis). ICRA5 moved to Tomar, Portugal, in 2013, in honor of Dr Lutz Edler, where actually the extension of Risk Analysis (RA) to Bioinformatics, Management and Industry was established. The Springer volume in 2013 provides the appropriate evidence.

Meanwhile ICRA6 moved to Barcelona, Spain, and it was jointly organized with the conference RISK2015. In 2017, the meeting ICRA7 was held in Chicago, in honor of Professor Ivette Gomes. One step forward, further from game theory, towards to more fields under risk, was offered by the second Springer volume, in 2018. ICRA8 took place in 2019, in honor of Professor Samad Hedayat and it was held in Vienna, Austria. Another step forward was given with the publication of the 3rd ICRA volume by Springer in 2022. This book provided an overview of the role of statistics in RA, by addressing theory, methodology and applications covering the broad scope of risk assessment in life sciences and public health, environmental science as well as in economics and finance. The conference also included as a main topic the Experimental Design, once it plays a key role in many of these areas.

ICRA8 brought together some of the most important researchers and practitioners working in both fields: Risk Analysis and Experimental Design.

ICRA9 took place in 2022 at Perugia, Italy. The extension of the subjects was even broader and the present volume, acting as a proceedings volume, reflects the situation. This meeting was supposed to take place in June 2021, but unfortunately due to COVID it was canceled at that time. A special thanks is addressed to the Local Organizing Committee, for being patient and determined to proceed, besides the restrictions and the pandemic impact all over the world.

In epidemiological studies, there is a need to identify and quantitatively assess the susceptibility of a portion of the population to specific risk factors. It is assumed that all the participants to the study have been equally exposed to the same possible hazard factors. The difference, at the early stage of the research study, is only due to a particular factor which acts as a susceptibility.

We, the people of ICRA, try all these years, with the conferences we organize in different places, to extend the field in scientific areas such as Food Science, Environmental Problems, Management, Economics, and Engineering, etc. Thanks to the interest of the distinguished participants, some improvements have been succeeded every time, in every conference. The Risk Analysis (RA) problem is not solely what in the Decision Theory, traditionally, is referred; in the early stages, it was also involved in political decisions. We had big and very profitable discussions, all the participants, at the early stages of ICRA conferences, where the Decision Analysis line of thought, in every meeting, was retreating and the supporters of such "solid line" were realizing the reality. Eventually, since Tomar conference in 2013, we have essentially improved the areas of application of RA.

Our thoughts are going to Dr Lutz Edler, DKFZ Germany, who was addressing successfully the stimulating and detailed discussions the first decade of the second millennium. Exchanging ideas remains the core value for ICRA, creative discussion can be always helpful to good will people—solid beliefs need a careful and detailed RA to be accepted. That is why we try to have a "general assembly meeting" at the end of the conference to discuss possible adjustments from the current ICRA to the next. We believe there is a further area for development: RA has been applied to different fields, where there is not the appropriate software to all the areas, such an example might be the area of Environmental Risk. As two of us have strong ties with ISI Committee on Risk Analysis (C. Kitsos, Chairman, Nov. 2013–Feb. 2015, T. Oliveira, Chairwoman, March 2015–today), our concern to RA and therefore to ICRA is a life target.

We deeply thank all the participants for the submitted papers. All the papers were reviewed by two independent reviewers, a tradition we try to keep, as it is a safe guide to keep our Quality Standards. We thank all the reviewers for their support to the final presentation of this volume. We are certain that all the participants enjoyed the hospitality in Perugia. Our sincere thanks are addressed to Springer, and especially to Dr Eva Hiripi, for the many years of excellent and kind cooperation. We are looking forward to see the ICRA10.

Egaleo, Greece Christos P. Kitsos
Lisboa, Portugal Teresa A. Oliveira
Perugia, Italy Francesca Pierri
Fisciano, Italy Marialuisa Restaino

Contents

Examining the Network Effects in Bank Risk: Evidence from Liquidity Creation in Mutual Banks

Carmelo Algeri, Antonio Fabio Forgione, and Carlo Migliardo

Abstract This study examines the impact of networking effects on the ability of Italian Credit Cooperative Banks (CCBs) to generate liquidity. The dynamics of this vital service that banks provide to the economy is conditioned, for CCBs, by the existence of spatial effects. Literature indicates that CCBs compete for a very similar local and niche market, giving rise to the hypothesis that changes in their credit and funding strategies have network-level effects. We conduct an empirical investigation using a suitable spatial model and controlling for potential endogeneity issues. Our hypothesis regarding the presence of spatial co-movement among CCB clusters is supported by the results. We provide evidence that spatial components exist, but that the geographical contemporaneous and non-contemporaneous terms balance each other out. We believe that the latter dependence is a result of the CCBs' ability to operate in a particular geographic area and serve similar clients. Our findings have implications for both the managerial decisions of CCBs and the policy actions that should take the reported effects into account.

Keywords Spatial dependence · Networking effects · Liquidity creation · Small banks

JEL Code R12, L14, G21, C33

C. Algeri (✉)
Department of Management, University of Bologna, Bologna, Italy
e-mail: carmelo.algeri@unibo.it

A. F. Forgione · C. Migliardo
Department of Economics, University of Messina, Messina, Italy
e-mail: fforgione@unime.it; cmigliardo@unime.it

© The Author(s), under exclusive license to Springer Nature Switzerland AG 2023
C. P. Kitsos et al. (eds.), *Statistical Modelling and Risk Analysis*,
Springer Proceedings in Mathematics & Statistics 430,
https://doi.org/10.1007/978-3-031-39864-3_1

1 Introduction

CCBs are a model of mutual and community banks in the strict sense. CCBs adopt "prevailing mutualism" rules, which entail: the adoption of the rule "one-head, one-vote"; that at least 70% of profits must be retained in a legal reserve, which cannot be distributed to members. In addition, shareholders are forbidden from holding shares with a total value exceeding 100,000 euros. The localism of CCBs derives primarily from the requirement that over half of risky activities must be conducted with members and that they direct at least 95% of their loan activity toward local agents, particularly small firms and households, who are typically also shareholders, with the strategic role of satisfying the credit demand of marginal customers, thereby reducing credit rationing issues (e.g., [1, 29]). Similarly, most of the CCB's funding comes from the surrounding community where the bank's headquarters are located.[1]

CCBs engage in traditional banking activities such as deposit-taking and loan granting; as a result, their assets are low-diversified and more vulnerable to liquidity risk, which is one of the main causes of bank failure, particularly if bank activity is highly skewed toward traditional operations [34]. Consequently, there is a significant incentive to generate liquidity, as it is strictly related to cost efficiency [12, 32] and bank profitability [31].[2] However, it is well known that banks reduce liquidity production in order to lower risk exposures and meet capital requirements imposed by bank regulations [42]. Similarly, the liquidity creation index is more effective and reliable than the conventional measure in identifying warning signals of bank fragility [24].

When an economic storm hits, whether in the form of idiosyncratic local shocks (even for a single bank) or systemic events, they are vulnerable to "bank run" in the form of fly-to-safety of customer-shareholders from weaker banks to stronger ones. In this framework, we believe that mutual banks operate in a closely related environment, forming a network in which they are affected by their neighbors' competitive activity. This geographical effect among CCBs could also regard liquidity creation, but very few studies have applied spatial models to investigate whether co-movement in bank activity is present. The identification of a

[1] The Italian cooperative banks were subjected to a significant reform (Law n. 49/2016) that, for the CCBs, was primarily aimed at enhancing their managerial and financial resources. CCBs were required to form a banking group directly under Single Supervisory Mechanism (SSM) and whose holding company is a stock company. CCBs must enter into a contract with their holding, which gives the latter the authority to set strategic guidelines for the CCBs, such as intervening to ensure compliance with the group's operational goals. As a result, CCBs continue to be cooperative and maintain their banking status, since the contract includes a coordinating authority. The Bank of Italy must approve the contracts; in particular, it regulates the holding's intervention authority over CCBs more stringently as CCBs become less financially sound. More details regarding the reform can be found at http://www.eacb.coop/en/news/members-news/bcc-thereform-of-the-co-operative-banks-in-italy-is-now-law.html.

[2] Even if they are non-profit organizations, mutual banks must be managed according to the principle of cost-effectiveness.

network in liquidity creation is important for several reasons. Indeed, the presence of network connectedness magnifies the relationship between liquidity creation and financial [23] and economic local stability [52]. However, this literature focuses only on emerging economies. Furthermore, our empirical specification allows us to investigate the direction and evolution of this neighboring effect over time, delving deeper into the spatial connection in mutual banks' ability to finance illiquid assets with liquid deposits. The findings show that neighboring CCBs have a significant impact on their liquidity creation activity, which we attribute to the well-defined market segment in which they typically compete.

After having introduced our spatial model and empirical specification in Sect. 2, we describe the data and variables adopted to carry out our analysis, together with the results of our empirical estimates (Sect. 3). Section 4 provides some concluding remarks.

2 Methodology

Our analysis is conducted following three main steps, namely the creation of the spatial terms, as detailed in Sect. 2.1, the estimation of bank market power, namely the Lerner index, which requires an estimation of the bank cost function, using the stochastic frontier analysis that we detail in the Appendix, and finally the estimation of the System GMM model, which allows us to investigate the relationship between bank liquidity creation and a set of determinants, also considering the spatial interacting effect (Sect. 2.2).

2.1 Spatial Model

Spatial econometrics evaluates spatial dependence effects [5], which, if present, can bias the estimation results of a standard regression model [38].[3]

Spatial autocorrelation,[4] namely spatial dependence and spatial heterogeneity (spatial non-stationarity), which are two components of spatial effects [4], describes a spatial structure that occurs when an observed value at location i is dependent on an observed value at location j.

The spatial econometric modeling employed in this paper relies on a Spatial Dynamic Panel Data (SDPD) model [33, 37, 39], which allows for spatial spillover effects among units to be accounted for. In detail, the Time-Space Dynamic (TSD) model [3] is considered, using the Generalized Method of Moments (GMM) dynamic estimator [18, 19, 21, 30, 48]. The spatial econometric model's general

[3] LeSage and Pace [41] argue that the traditional regression model's estimated parameters are incorrect and inconsistent if the independence hypothesis between observations is violated.

[4] Typically, spatial autocorrelation refers to a weaker form of spatial dependence [3].

form is as follows:

$$y_{i,t} = \alpha y_{i,t-1} + \rho \sum_{j \neq i} \boldsymbol{w}_{ij} \cdot y_{j,t} + \lambda \sum_{j \neq i} \boldsymbol{w}_{ij} \cdot y_{j,t-1} + \kappa \mathbf{x}_{i,t} + (\upsilon_i + \varepsilon_{i,t}) \quad (1)$$

$$|\alpha| < 1, \ |\rho| < 1, \ |\lambda| < 1; \ \ i = 1 \ldots N; \ \ t = 1 \ldots T$$

where $\sum_{j \neq i} \boldsymbol{w}_{ij} \cdot y_{j,t}$ $\left(\text{or} \sum_{j \neq i} \boldsymbol{w}_{ij} \cdot y_{j,t-1}\right)$ is the sum of the dependent variable (lagged dependent variable) of all other units j, weighted by the elements of the spatial weight matrix w_{ij}, and represents the degree of interconnectedness between units i and j, $\mathbf{x}_{i,t}$ is the vector of explanatory variables, and υ_i and $\varepsilon_{i,t}$ are the two elements that make up the error term.

The unknown coefficients to be estimated are denoted by the terms α, ρ, λ, and κ, which refer to serial dependence, spatial dependence, spatial-temporal dependence, and control variables, respectively. The sum of the first three terms must be less than one to satisfy the condition of global stationarity (i.e., $|\alpha + \rho + \lambda| < 1$), otherwise the dependent variable has to be appropriately treated to remove a possible unit root issue.[5]

The distance spatial weights matrix (SWM), denoted by $\boldsymbol{W}$, is composed of weights ($w_{ij} \ : \ i, \ j \ = \ 1, \ldots, n$) that are transformations of the initial distances becoming zero on the principal diagonal ($w_{ii} \ = \ 0, \ \forall i$), since no spatial unit is a neighbor of itself.

We consider three distinct types of SWMs to provide a comprehensive overview of the potential spatial co-movement in the CCBs liquidity creation strategy: a distance matrix with a distance cut-off parameter of 52 km, a matrix resulting from multiplying the latter distance matrix by a quadratic (or Epanechnikov) kernel function, and an inverse distance squared matrix.[6]

In more detail, the distance matrix contains elements (w_{ij}) that takes the value 1 if the distance d_{ij} between two spatial units i and j is less than the distance cut-off (52 km), and 0 otherwise. Following [8], we have determined the quadratic SWM as follows:

$$w_{ij} = (3/4)(1 - z^2) \text{ for } |z| < 1$$

where z is the distance d_{ij} from the geographical units i to j and the bandwidth h_i. This ensures that z is always less than 1. As explained above, we have multiplied the quadratic matrix for distance matrix. Finally, the inverse distance power SWM employs a continuous parameterized function of distances such that distance decreases with increasing distance according to a negative exponential ($w_{ij} = e^{-\beta d_{ij}}$), which we set to 2 (for more detail, see [8]).

---

[5] For further information on estimating the unit root in the SDPD model, see [51].

[6] Specifically, the spatial matrices are obtained with GeoDa software (for further details, see [6]) and then used in Stata.

2.2 Econometric Specification

To give accurate estimates and solve endogeneity concerns, we consider the System GMM model (SYS-GMM) [9, 17].[7] All explanatory variables are time-lagged.

The following is the TSD model's empirical equation:

$$\begin{aligned} NLC_{i,t} &= \xi + \beta NLC_{i,t-1} + \varrho \sum_{j \neq i} \boldsymbol{w}_{ij} \cdot NLC_{j,t} + \delta \sum_{j \neq i} \boldsymbol{w}_{ij} \cdot NLC_{j,t-1} \\ &+ \vartheta Lerner_{i,t-1} + \tau Size_{i,t-1} + \varsigma Tier\ 1_{i,t-1} + \phi Int_{i,t-1} \\ &+ \varphi Ln[Crime]_{i,t-1} + \theta Sanction_i + (\upsilon_i + \epsilon_{i,t}) \end{aligned} \tag{2}$$

NLC is the variable adopted as a proxy of liquidity creation as in [14], but excludes the off-balance-sheet activities due to their marginal role in CCBs. *NLC* is a ratio calculated by comparing bank weight assets and bank weight liabilities and equity in the numerator and denominator, respectively. The assets and liabilities of the bank are classified as liquid, semi-liquid, or illiquid based on how simple, costly, and timely it is to convert them to liquid funds. In detail, the liquid assets are the sum of cash and deposits to the bank, while the liquid liabilities are the bank and customers' deposits, which are weighted with a coefficient of -0.5 and 0.5, respectively. The illiquid assets of the bank are its fixed assets, while the illiquid liabilities are its subordinate debt, other liabilities, and bank equity. Each of these items has a weight of ± 0.5, respectively. Both semi-liquid assets, such as bank loans and financial assets, and semi-liquid liabilities, including debt securities, trading liabilities, and derivatives, account for zero. Since our data does not permit us to differentiate between commercial and non-commercial loans (considered more illiquid), we adopt a more conservative measure of bank liquidity creation by including all bank loans under the category with a zero coefficient.

It has been proved that bank market power is a factor affecting liquidity creation, but the sign of the associated coefficient is sensitive to the competition proxy variable adopted. Specifically, the Herfindahl-Hirschman index, which measures market share concentration is directly related to the liquidity variable [22, 36], while the Lerner index, proposed by [40], captures the capacity to practice a markup over the marginal costs and is inversely related to the liquidity variable (e.g., [15, 26, 35]). The Lerner index has a clear benefit over other competition metrics in that it provides bank-level estimates of competition [26]. Furthermore, we believe that the Lerner index provides a more accurate picture of market power for banks operating in niche markets than the cooperative, which typically holds a low market share but can charge higher markups to customers who lack access to alternative funding channels. Therefore, the Lerner index, as stated, measures a firm's ability to charge

[7] The GMM models are calculated using [50] finite sample correction and FOD transformation.

prices higher than its marginal costs. We report in Appendix more technical details regarding the estimation of our *Lerner* variable.

The log of the total assets (*Size*), serves as a measure of the size of the bank. Actually, well-established literature points out that size could affect a bank's liquidity creation ability indirectly [27, 46] on the assumption that large banks diversify their assets more.

Bank capitalization, denoted with *Tier 1*, is the buffer used to hedge against liquidity mismatches in bank activity, and [14] developed two opposing hypotheses to identify the possible effect of this variable. According to the "financial fragility-crowding out hypothesis", a low capital buffer makes a bank vulnerable and leads to increased monitoring activity, which leads to more loans being granted. Furthermore, for bank activity, equity is preferred over deposits. According to the "risk absorption hypothesis", bank capital serves as a hedge against bank risk. As a result, higher capital shares promote better bank liquidity transformation. Berger et al. [14] show that large banks confirm the former assumption, while small banks validate the latter. However, the empirical literature is still controversial (for a comprehensive review, see [36]) and thereby we follow an agnostic view for this variable.

There are no studies relating the impact of crime to the creation of liquidity, and there are few cross-country analyses based on national indicators of the rule of law that confirm the direct impact of a favorable institutional context on bank activity (e.g., [16]). To check for this potential spillover effect, we adopted an indicator of crime presence (*Crime*) provided by Istat, which represents the crimes reported to the judicial authorities by the police forces and is calculated at the provincial level.[8]

The specification includes *Int*, which indicates, at provincial level, the lending rates on revocable loans.[9] When interest rates are relatively high, bank risk management implements prudential credit policy, financing only the most creditworthiness clients, due to the increased likelihood of encountering adverse selection and moral hazard issues. In addition, the negative effects of high interest rates extend to the funding of the bank. Indeed, higher interest rates are well known to encourage longer-term, less liquid investments, crowding out the demand for bank deposits. Therefore, interest rate should be another factor curbing bank liquidity creation.

Finally, we use a dummy variable (*Sanction*) equal to one for years in which the CCB has been subjected to a disciplinary procedures by bank authority as in [47]. Indeed, small banks can be sanctioned by the central bank for a variety of reasons, but the end result is that bank operations are limited.

[8] Data is available on the Istat website https://esploradati.istat.it/databrowser/#/en.

[9] Data has been provided by Bank of Italy (table TRI30830) https://infostat.bancaditalia.it/.

Table 1 Summary statistics

Variable	Mean	Std. Dev.	Q1	Median	Q3	Min	Max
NLC	0.452	0.060	0.416	0.456	0.490	0.182	0.676
Lerner	0.378	0.113	0.304	0.379	0.452	0.000	0.630
Size	12.961	1.020	12.151	13.043	13.634	9.552	16.281
Tier 1	0.192	0.080	0.141	0.172	0.222	0.065	0.798
Int	0.071	0.020	0.054	0.072	0.084	0.038	0.111
Ln[*Crime*]	8.203	0.230	8.053	8.155	8.319	7.685	9.046
Sanction	0.131	0.340	0	0	0	0	1

The number of observations is 1518 for all the variables

Table 2 Correlation matrix for the data shown in Table 1

	1.	2.	3.	4.	5.	6.	7.
1. *NLC*	1						
2. *Lerner*	−0.271*	1					
3. *Size*	0.194*	0.287*	1				
4. *Tier* 1	−0.388*	−0.038	−0.435*	1			
5. *Int*	−0.363*	0.086*	−0.169*	0.095*	1		
6. *Ln*[*Crime*]	−0.079*	0.099*	0.107*	−0.036	0.183*	1	
7. *Sanction*	0.006	−0.022	0.063*	−0.112*	0.106*	−0.042	1

* Significance at the 5% level or lower

3 Data and Results

This study uses a sample of 253 Italian CCBs over the period 2011–2017, for a total of 2024 observations in a balanced panel.[10] This is a sub-sample of CCBs because we removed those without financial data available for all years of interest, as well as the six isolated banks.[11]

The accounting data comes from the Bureau van Dijk Orbis-Bank Focus (BvD Orbis) database[12] and provincial macroeconomic indicators are from ISTAT and Bank of Italy.

Tables 1 and 2 report the summary statistics and the correlation matrix between the variables adopted in our empirical model.

It is important to note that the *NLC* of CCBs is lower than one, because we use a conservative measure of liquidity creation, which excludes bank loans.

Each CCB headquarters was geo-referenced in terms of geographical coordinates to create the geospatial dataset. Figure 1 depicts the spatial network linking all CCBs' headquarters based on our spatial matrices.

---

[10] The SDPD model necessitates a well-balanced panel.

[11] Sardinia had two CCBs omitted, as did Elba Island, Aosta Valley, Sicily, and Puglia.

[12] Because certain data are missing, we acquired them from the CCBs' websites.

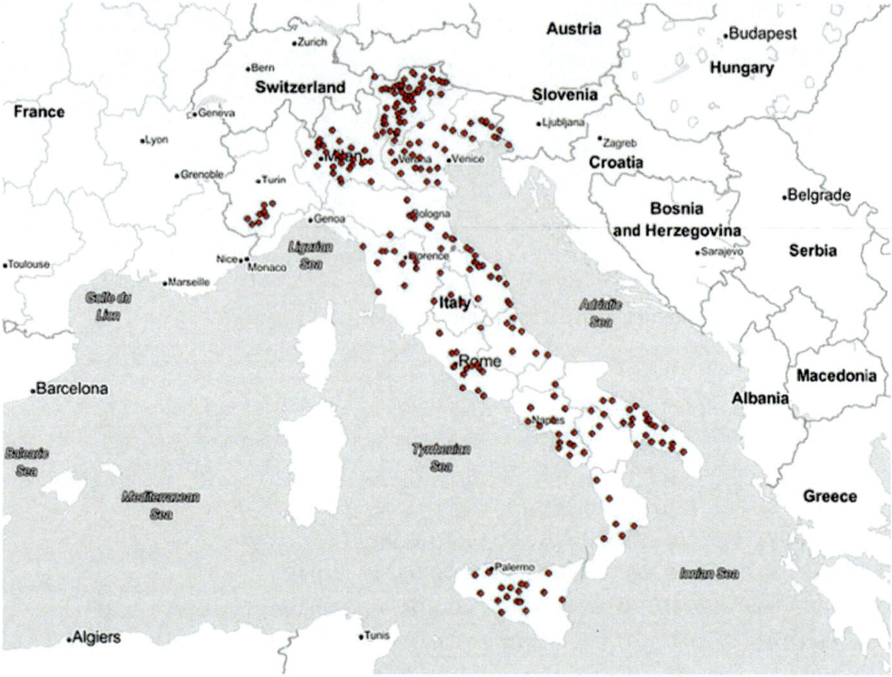

Fig. 1 Spatial distribution of Italian cooperative banks

The map in Fig. 1 shows that there are numerous places in Italy with a high concentration of CCBs, which supports our goal of investigating their spatial correlation more thoroughly.

Our hypothesis regarding the occurrence of co-movement in the liquidity creation ability of CCBs is supported by three Lagrange Multiplier (LM) tests, which are shown in Table 3.

In detail, the test results confirm the presence of spatial connections [5, 20]. Similarly, the joint and conditional LM tests [10] validate the previous outcome. Finally, the [11] tests confirm the existence of spatial interconnections as well as serial correlation.

Numerous studies have been conducted to model the bank risk dynamics because this intermediary constitutes one of the most important and vulnerable factors in modern economies. The present analysis follows the strand of literature that adopts liquidity creation as a measure of bank risk.

Table 4 reports the results of our empirical estimates.

Across all estimated models, the positive coefficients of the auto-regressive variables indicate that the decisions regarding the CCBs' liquidity creation strategy have a lasting impact. Thus, the bank's decision to increase or decrease liquidity is repeated the following year. The coefficients associated with spatial lag and time-lag variables have positive and negative values, respectively. The combined effect of

Table 3 LM tests for spatial, serial correlation and random effects

LM test description	Statistic	P-value
Anselin [2]		
Conditional test for spatial error autocorrelation		
(H_0: spatial error autoregressive coefficient equal to zero)	8.91	0.000
Conditional test for spatial lag autocorrelation		
(H_0: spatial lag autoregressive coefficient equal to zero)	28.85	0.000
Baltagi et al. [10]		
Joint test		
(H_0: absence of random effects and spatial autocorrelation)	1338.2	0.000
Marginal test of random effects		
(H_0: absence of random effects)	30.76	0.000
Marginal test of spatial autocorrelation		
(H_0: absence of spatial autocorrelation)	19.79	0.000
Conditional test of spatial autocorrelation		
(H_0: absence of spatial autocorrelation, assuming random effects are non null)	24.49	0.000
Conditional test of random effects		
(H_0: absence of random effects, assuming spatial autocorrelation may or may not be equal to 0)	37.40	0.000
Baltagi et al. [11]		
Joint test		
(H_0: absence of serial or spatial error correlation or random effects)	1404.7	0.000
One-dimensional conditional test		
(H_0: absence of spatial error correlation, assuming the existence of both serial correlation and random effects)	199.16	0.000
One-dimensional conditional test		
(H_0: absence of serial correlation, assuming the existence of both spatial error correlation and random effects)	129.33	0.000
One-dimensional conditional test		
(H_0: absence of random effects, assuming the existence of both serial and spatial error correlation)	119.53	0.000

the two coefficients, mainly for two of the three spatial models, makes the network effect on the liquidity creation of CCBs within the considered time frame almost insignificant. The consistency appears to be due to the fact that these banks target clients with similar economic characteristics and geographic locations, implying that both aggressive commercial strategies and even macroeconomic shocks are offset over time. Consequently, the primary function of a bank tends to remain stable across time within the CCB community.

CCBs that can charge prices above their marginal costs generate less liquidity, and thus competition is a factor that impedes bank liquidity creation, most likely as a result of a reduction in demand-side liquidity [36]. Market dominance permits

Table 4 Estimates of TSD model

	Non-spatial model	Spatial dynamic models		
		W_1	W_2	W_3
NLC	(1)	(2)	(3)	(4)
NLC_{t-1}	0.5415*** (0.103)	0.5568** (0.234)	0.6232*** (0.155)	0.6759*** (0.180)
$W \times NLC_t$		0.7098*** (0.259)	0.6350*** (0.173)	0.7913*** (0.258)
$W \times NLC_{t-1}$		−0.7771** (0.312)	−0.5886*** (0.184)	−0.8810*** (0.257)
$Lerner$	−0.2289*** (0.056)	−0.1227** (0.062)	−0.1469** (0.072)	−0.2107*** (0.074)
$Size$	−0.0149** (0.007)	−0.0158* (0.009)	−0.0256** (0.012)	−0.0245** (0.011)
$Tier$ 1	−0.3731*** (0.097)	−0.2924** (0.116)	−0.3465*** (0.116)	−0.3071** (0.129)
Int	−0.5044*** (0.168)	−0.6347** (0.291)	−0.4171** (0.208)	−0.4527** (0.198)
$Ln[Crime]$	0.0077 (0.009)	0.0138 (0.009)	0.0196 (0.012)	0.0318 (0.039)
$Sanction$	−0.0116** (0.005)	−0.0094** (0.004)	−0.0112** (0.005)	−0.0182** (0.009)
No. Instruments	25	31	31	31
AR(1)	0.0000	0.0020	0.0000	0.0000
AR(2)	0.3268	0.2078	0.3759	0.3183
Hansen test	0.2536	0.1550	0.5919	0.3254
CD (Pesaran [44][a])	4.37***	−0.99	−0.36	−0.02

Robust standard errors are reported in parentheses [50]. Year dummies and constant term in all regressions. The estimates regard 253 banks for a total of 1518 observations. Pesaran [45]: 5.39 (P-value < 0.01)

* $p < 0.1$

** $p < 0.05$

*** $p < 0.01$

[a] CD (Pesaran [44]) is not a TSD model test, but it is necessary to evaluate the presence of *ex-ante* and *ex-post* strong cross-sectional dependence

CCBs to cherry-pick when selecting and monitoring borrowers, denying credit and cutting off loans to less creditworthy borrowers.

The *Size* estimates support the indirect effect that bank size has on liquidity creation, which has been well documented in the literature. This effect also applies to small banks, such as CCBs. More bank capital (*Tier 1*) has also the effect of lowering the liquidity produced by CCBs, supporting the "financial fragility-crowding out" hypothesis, which applies perfectly to the mutual banks framework. Indeed, bank law limits CCB shareholders to small investors primarily interested in better bank service conditions and moreover living in the bank's operating area. Given this management constraint, CCBs with less capital tend to expand their lending activity, and, with all things being equal, create more liquidity, in order to retain and attract shareholders, who are frequently looking for better credit terms.

Both interest rates and being sanctioned are negatively related to our dependent variable, as expected. Instead, the bank's capacity to generate liquidity is unaffected by the level of provincial crime.

The goodness of our non-spatial and SDPD models are confirmed by the few instruments used (20 and 28, respectively) with respect to the number of CCBs in the panel (253), as well as by the Arellano-Bond first and second-order autocorrelation tests (AR1 and AR2), which confirm the adequateness of our GMM estimates. Similarly, the Hansen test validates the appropriateness of the instrumental variables employed. Finally, the estimated coefficients β ϱ and δ, add up to less than one, excluding the presence of a unit root problem in the estimates.

Table 4 also reports the post-estimation test on the residuals of the SPDP models [44] to verify that the technique solves the cross-section correlation in the estimate. The test result confirms the assertion that spatial model errors are cross-sectionally independent. We stress-tested our model to evaluate its predictive performance. In particular, we evaluated the out-of-sample prediction performance of our model, by excluding the first three time periods (2011–2013) from the estimation sample for each bank in the panel. The routine then recursively fitted the specified models to the remaining subsample and used the resulting parameters to forecast the dependent variable in the unused periods (out-of-sample). The estimated root square mean error is quite low (0.6). We also adopted alternative periods, with similar results in terms of RSME (see, [49]).[13]

4 Concluding Remarks

The recent economic crisis has highlighted the significance of the credit crunch, which is both a cause and a result of the major advanced economies' economic downturn. In a scenario in which disintermediation appears to be widespread for banks, we investigated the liquidity creation determinants for a type of intermediary that operates with small borrowers (families and micro and small enterprises) that frequently rely on bank credit as their only source of funds. In this framework, we demonstrated how the network of CCBs located in the same territory influences maturity transformation activity.

The implications of this regularity are significant, not only on a financial level but also in terms of economic repercussions for regions of Italy where credit rationing issues are more prevalent, such as the less developed Italian regions. Similarly, the transmission of the monetary policy mechanism is altered as a result of a money tightening, causing negative externalities that exacerbates the credit crunch phenomenon as a result of the fly-to-safety strategy. The spatial effect exhibiting the opposite sign over time is a further indicator that has a noticeable impact on financial stability. This implies that each idiosyncratic shock increases the fluctuation of the liquidity creation dynamic.

The model provides a useful relationship to indirectly evaluate the bank's financial distress. In fact, our findings support the "financial fragility-crowding out"

[13] An anonymous referee suggested this robustness check.

hypothesis, so high liquidity creation is associated with a low capital buffer, which in turn increases the likelihood of bank distress. Finally, it should be kept in mind that this study is an initial attempt to assess the importance of spillover in bank liquidity transformation, and that additional research is required to validate the geographic effect in other types of local banks, such as commercial banks. Further avenues of investigation are made possible using our empirical model, which, even if it incorporates a set of control variables, requires a small number of instruments, and allows further improvement by incorporating additional variables that may influence bank liquidity creation.

Appendix

As described in the manuscript, the Lerner index is calculated as follows:

$$Lerner_{it} = \frac{(P_{it} - MC_{it})}{P_{it}} \tag{3}$$

where P are the bank's prices, proxied by the ratio of the sum of interest and non-interest income to total assets and MC its marginal costs.

To estimate the marginal costs of CCBs, it is necessary to specify a cost function, such as a transcendental logarithmic cost function, relating bank total costs ($C_{i,t}$—in our specification equal to total operating expense) to bank output ($Q_{i,t}$) that we proxy to total assets (see, for instance, [13, 25, 28, 43]) and bank inputs. Specifically, we consider the price of bank inputs: P_1—personnel expenses over total assets (labor price), P_2—other administrative and operating expenses divided by fixed assets (fixed capital price) and P_3—interest expenses over bank funding (borrowed funds cost). In line with previous studies on this topic (e.g., [13, 25, 28, 43]) Therefore, the total costs are determined as follows:

$$\begin{aligned} lnC_{i,t} &= \alpha_0 + \alpha_1 lnQ_{i,t} + \frac{1}{2}\alpha_2 lnQ_{i,t}^2 + \sum_{k=1}^{3} \beta_k lnP_{k,i,t} \\ &+ \frac{1}{2}\sum_{k=1}^{3}\sum_{j=1}^{3} \gamma_{k,j} lnP_{k,i,t} lnP_{j,i,t} + \frac{1}{2}\sum_{k=1}^{3} \zeta_k lnQ_{i,t} lnP_{k,i,t} \\ &+ (\upsilon_{i,t} + \nu_{i,t}) \end{aligned} \tag{4}$$

$\upsilon_{i,t}$ captures the actual cost inefficiency term and is structured as a truncated non-negative random variable $N^+(0, \sigma_u^2)$, whereas $\nu_{i,t}$ represents the white noise independent and identically distributed with 0 mean and variance σ_v^2.

We imposed the linear homogeneity condition in input prices as follows:

$$\begin{cases} \sum \beta_k = 1 \; \forall k \\ \sum \gamma_h = 0 \; \forall h \\ \sum \zeta_j = 0 \; \forall j \end{cases}$$

After having estimated the translog cost function by applying the stochastic frontier analysis, we use the coefficients $\hat{\alpha}_1$, $\hat{\alpha}_2$, $\hat{\zeta}_1$, $\hat{\zeta}_2$, $\hat{\zeta}_3$, to determine the bank marginal costs by calculating the partial derivative of Eq. (4) with respect to Q. In detail:

$$MC_{i,t} = \frac{\partial C_{i,t}}{\partial Q_{i,t}} = \frac{\partial lnC_{i,t}}{\partial lnQ_{i,t}} \frac{C_{i,t}}{Q_{i,t}} = \left(\hat{\alpha}_1 + \hat{\alpha}_2 \, lnQ_{i,t} + \sum_{k=1}^{3} \hat{\zeta}_k \, lnP_{k,i,t} \right) \frac{C_{i,t}}{Q_{i,t}} \tag{5}$$

References

1. Aiello, F., Bonanno, G.: Multilevel empirics for small banks in local markets. Pap. Reg. Sci. **97**(4), 1017–1037 (2018)
2. Anselin, L.: Spatial econometrics: methods and models. Vol. 4. Springer Science & Business Media (1988)
3. Anselin, L.: Spatial econometrics. In BH Baltagi (Ed.): A Companion to Theoretical Econometrics, pp. 310–330. Blackwell Publishing: Malden (2001)
4. Anselin, L.: Thirty years of spatial econometrics. Pap. Reg. Sci. **89**(1), 3–25 (2010)
5. Anselin, L.: Spatial Econometrics: Methods and Models, vol. 4. Springer Science & Business Media, Berlin (2013)
6. Anselin, L., Rey, S.J.: Modern Spatial Econometrics in Practice: A Guide to GeoDa, GeoDaSpace and PySAL. GeoDa Press LLC, Chicago (2014)
7. Anselin, L., Le Gallo, L., Jayet, H.: Spatial panel econometrics. In: Matyas, L., Sevestre, P. (eds.) The Econometrics of Panel Data Fundamentals and Recents Developments in Theory and Practice, pp. 625–660. Berlin, Heidelberg: Springer Berlin Heidelberg (2008)
8. Anselin, L., Syabri, I., Kho, Y.: Geoda: an introduction to spatial data analysis. In: Handbook of Applied Spatial Analysis, pp. 73–89. Springer, Berlin (2010)
9. Arellano, M., Bover, O.: Another look at the instrumental variable estimation of error-components models. J. Econ. **68**(1), 29–51 (1995)
10. Baltagi, B.H., Song, S.H., Koh, W.: Testing panel data regression models with spatial error correlation. J. Econ. **117**(1), 123–150 (2003)
11. Baltagi, B., Song, S., Jung, B., Koh, W.: Testing panel data regression models with spatial and serial error correlation. J. Econ. **140**, 5–51 (2007)
12. Baltas, K.N., Kapetanios, G., Tsionas, E., Izzeldin, M.: Liquidity creation through efficient M&As: a viable solution for vulnerable banking systems? Evidence from a stress test under a panel VAR methodology. J. Bank. Financ. **83**, 36–56 (2017)
13. Beck, T., De Jonghe, O., Schepens, G.: Bank competition and stability: cross-country heterogeneity. J. Financ. Intermed. **22**(2), 218–244 (2013)
14. Berger, A.N., Klapper, L.F., Turk-Ariss, R.: Bank competition and financial stability. J. Financ. Serv. Res. **35**(2), 99–118 (2009)

15. Berger, A.N., Klapper, L.F., Turk-Ariss, R.: Bank competition and financial stability. In: Handbook of Competition in Banking and Finance. Edward Elgar Publishing, Cheltenham (2017)
16. Berger, A.N., Li, X., Saheruddin, H., Zhao, D.: Government guarantees and bank liquidity creation around the world. Available at SSRN 3729115 (2020)
17. Blundell, R., Bond, S.: Initial conditions and moment restrictions in dynamic panel data models. J. Econ. **87**(1), 115–143 (1998)
18. Bouayad-Agha, S., Védrine, L.: Estimation strategies for a spatial dynamic panel using GMM. A new approach to the convergence issue of European regions. Spat. Econ. Anal. **5**(2), 205–227 (2010)
19. Bouayad-Agha, S., Turpin, N., Védrine, L.: Fostering the development of European regions: a spatial dynamic panel data analysis of the impact of cohesion policy. Reg. Stud. **47**(9), 1573–1593 (2013)
20. Breusch, T.S., Pagan, A.R.: The Lagrange multiplier test and its applications to model specification in econometrics. Rev. Econ. Stud. **47**(1), 239–253 (1980)
21. Cainelli, G., Montresor, S., Marzetti, G.V.: Spatial agglomeration and firm exit: a spatial dynamic analysis for Italian provinces. Small Bus. Econ. **43**(1), 213–228 (2014)
22. Casu, B., Di Pietro, F., Trujillo-Ponce, A.: Liquidity creation and bank capital. J. Financ. Serv. Res. **56**(3), 307–340 (2019)
23. Chen, T.H., Lee, C.C.: Spatial analysis of liquidity risk in China. N. Am. J. Econ. Financ. **54**, 101047 (2020)
24. Chen, T.H., Lee, C.C., Shen, C.H.: Liquidity indicators, early warning signals in banks, and financial crises. N. Am. J. Econ. Financ. **62**, 101732 (2022)
25. Coccorese, P., Ferri, G.: Are mergers among cooperative banks worth a dime? Evidence on efficiency effects of M&As in Italy. Econ. Model. **84**, 147–164 (2020)
26. Dang, V.D.: Bank funding, market power, and the bank liquidity creation channel of monetary policy. Res. Int. Bus. Financ. **59**, 101531 (2022)
27. Dang, V.D., et al.: How do bank characteristics affect the bank liquidity creation channel of monetary policy? Financ. Res. Lett. **43**, 101984 (2021)
28. Danisman, G.O., Demirel, P.: Bank risk-taking in developed countries: the influence of market power and bank regulations. J. Int. Financ. Mark. Inst. Money **59**, 202–217 (2019)
29. Destefanis, S., Barra, C., Lubrano Lavadera, G.: Financial development and local growth: evidence from highly disaggregated Italian data. Appl. Financ. Econ. **24**(24), 1605–1615 (2014)
30. Donfouet, H.P.P., Jeanty, P.W., Malin, E.: Analysing spatial spillovers in corruption: a dynamic spatial panel data approach. Pap. Reg. Sci. **97**, S63–S78 (2018)
31. Duan, Y., Niu, J.: Liquidity creation and bank profitability. N. Am. J. Econ. Financ. **54**, 101250 (2020)
32. Duan, Y., Fan, X., Li, X., Rong, Y., Shi, B.: Do efficient banks create more liquidity: international evidence. Financ. Res. Lett. **42**, 101919 (2021)
33. Elhorst, J.P.: Spatial Econometrics: From Cross-Sectional Data to Spatial Panels, vol. 479. Springer, Berlin (2014)
34. Fungacova, Z., Turk, R., Weill, L.: High liquidity creation and bank failures. J. Financ. Stab. **57**, 100937 (2021)
35. Horvath, R., Seidler, J., Weill, L.: How bank competition influences liquidity creation. Econ. Model. **52**, 155–161 (2016)
36. Hsieh, M.F., Lee, C.C., Lin, Y.C.: New evidence on liquidity creation and bank capital: the roles of liquidity and political risk. Econ. Anal. Policy **73**, 778–794 (2022)
37. Jeong, H., Lee, L.F.: Spatial dynamic models with intertemporal optimization: specification and estimation. J. Econ. **218**, 82–104 (2020)
38. Kopczewska, K., Lewandowska, A.: The price for subway access: spatial econometric modelling of office rental rates in London. Urban Geogr. **39**(10), 1528–1554 (2018)
39. Lee, L.f., Yu, J.: A spatial dynamic panel data model with both time and individual fixed effects. Economet. Theor. **26**(2), 564–597 (2010)

40. Lerner, A.P.: Economic theory and socialist economy. Rev. Econ. Stud. **2**(1), 51–61 (1934)
41. LeSage, J., Pace, R.K.: Introduction to Spatial Econometrics. Chapman and Hall/CRC, New York (2009)
42. Nguyen, T.V.H., Ahmed, S., Chevapatrakul, T., Onali, E.: Do stress tests affect bank liquidity creation? J. Corp. Finan. **64**, 101622 (2020)
43. Okolelova, I., Bikker, J.A.: The single supervisory mechanism: competitive implications for the banking sectors in the euro area. Int. J. Financ. Econ. **27**(2), 1818–1835 (2022)
44. Pesaran, M.H.: General diagnostic tests for cross-sectional dependence in panels. University of Cambridge, Cambridge Working Papers in Economics, vol. 435 (2004)
45. Pesaran, M. H.: Testing weak cross-sectional dependence in large panels. Econometric Reviews, **34**(6–10), 1089–1117 (2015).
46. Pham, H.S.T., Le, T., Nguyen, L.Q.T.: Monetary policy and bank liquidity creation: does bank size matter? Int. Econ. J. **35**(2), 205–222 (2021)
47. Roman, R.A.: Winners and losers from supervisory enforcement actions against banks. J. Corp. Finan. **60**, 101516 (2020)
48. Segura III, J.: The effect of state and local taxes on economic growth: a spatial dynamic panel approach. Pap. Reg. Sci. **96**(3), 627–645 (2017)
49. Ugarte-Ruiz, A.: Xtoos: stata module for evaluating the out-of-sample prediction performance of panel-data models. Statistical Software Components S458710, Boston College Department of Economics (2023)
50. Windmeijer, F.: A finite sample correction for the variance of linear efficient two-step GMM estimators. J. Econ. **126**(1), 25–51 (2005)
51. Yu, J., Lee, L.F.: Estimation of unit root spatial dynamic panel data models. Econ. Theory **26**(5), 1332–1362 (2010)
52. Zhang, X., Fu, Q., Lu, L., Wang, Q., Zhang, S.: Bank liquidity creation, network contagion and systemic risk: evidence from Chinese listed banks. J. Financ. Stab. **53**, 100844 (2021)

Teaching Note—Data Science Training for Finance and Risk Analysis: A Pedagogical Approach with Integrating Online Platforms

Afshin Ashofteh

Abstract The main discussion of this paper is a method of data science training, which allows responding to the complex challenges of finance and risk analysis. There is growing recognition of the importance of creating and deploying financial models for risk management, incorporating new data and Big Data sources. Automating, analyzing, and optimizing a set of complex financial systems requires a wide range of skills and competencies that are rarely taught in typical finance and econometrics courses. Adopting these technologies for financial problems necessitates new skills and knowledge about processes, quality assurance frameworks, technologies, security needs, privacy, and legal issues. This paper discusses a pedagogical approach to overcome the teaching complexity of needed soft and hard skills in an integrated manner with its advantages, disadvantages, and vulnerabilities.

Keywords Data science · Finance · Risk · Pedagogical · Active learning

1 Introduction

Data science in finance and risk analysis is an analytical ability to function effectively in financial markets where financial data are analyzed to make decisions. Data science for finance brings a range of thinking and practical skills. It includes foundations in mathematics, statistics, computer science, and finance. Moreover, the sensitivity of the outcomes to data quality needs data engineering skills [1].

In finance and risk management, the capacity to incorporate new and Big Data sources [2] and benefit from emerging technologies are investigated by

A. Ashofteh (✉)
NOVA Information Management School (NOVA IMS), Universidade Nova de Lisboa, Lisboa, Portugal
e-mail: aashofteh@novaims.unl.pt
https://novaresearch.unl.pt/en/persons/afshin-ashofteh

© The Author(s), under exclusive license to Springer Nature Switzerland AG 2023
C. P. Kitsos et al. (eds.), *Statistical Modelling and Risk Analysis*,
Springer Proceedings in Mathematics & Statistics 430,
https://doi.org/10.1007/978-3-031-39864-3_2

many scholars and big consultancy companies [3]. Web technologies, remote data collection techniques, user experience platforms, and blockchain brings new fields of knowledge and competencies in finance, which are necessary to automate, analyze, and optimize complex financial systems. These new scientific paradigms of information and knowledge are not included in most traditional courses in finance and econometrics [4]. It necessitates new knowledge and skills [5] and new teaching approaches to empower those thinking about a career in data science for finance to upgrade with new quality assurance frameworks, technologies, and even legislations in security, privacy, and ethical issues [6]. This implies that financial data scientists should be aware of the data protection regulations, common ethical concerns arising in financial activities, and relevant ethical guidance by national, regional, and international regulators such as central banks, and the securities and market authorities [7].

However, there are methodological issues and debates among academics concerning the many constraints to teaching the vast range of hard and soft skills and abilities considered necessary for teaching data science in finance and risk management [8]. Learning the necessary skills to use the data science solutions for financial problems effectively requires learners to be more proactive and analytically literate. Financial data scientists need to make data, methods, and outcomes more intelligible to end users [9]. They should be enabled to mature the financial risks analysis and apply common sense to problems to extract timely relevant information considering the risk and uncertainty attached to them. There is a need for a course in the field of financial data science to focus on these specific needs and issues relevant to financial activities and risk management. This paper develops a framework of the essential elements for active learning of financial data science to form a meaningful learning experience. First, it presents a graphical summary in Fig. 1 to show the role of data science in finance and risk management business processes. Figure 1 shows the model consists of (1) Building methodology and related theories in two design and build phases; (2) integrating methodologies with data engineering by data curators; (3) extracting the strategies based on sustainable algorithms, which are the result of combining machine learning and methodologies; (4) meta-strategy and backtesting and approval of the investment committee; (5) Graduation phase for automation and industrialization to build intelligent systems; and finally; (6) deploying the result to the platforms and checking the strategies for re-allocation if necessary. As we can see in Fig. 1, soft skills such as financial thinking, data and statistical literacy, and specific knowledge of ethical codes, regulations, and dissemination of financial information are critical requirements of data science in finance.

Second, the paper analyses and discusses the educational requirements of this model, clarifying their contribution, interactions, and current and future importance in analysing the financial data. The learning method combines different online platforms in a harmonized way to develop soft and hard skills about data science for finance and its scientific paradigms. As a result, it provides information on the structure and content of a course about financial data science with an active learning

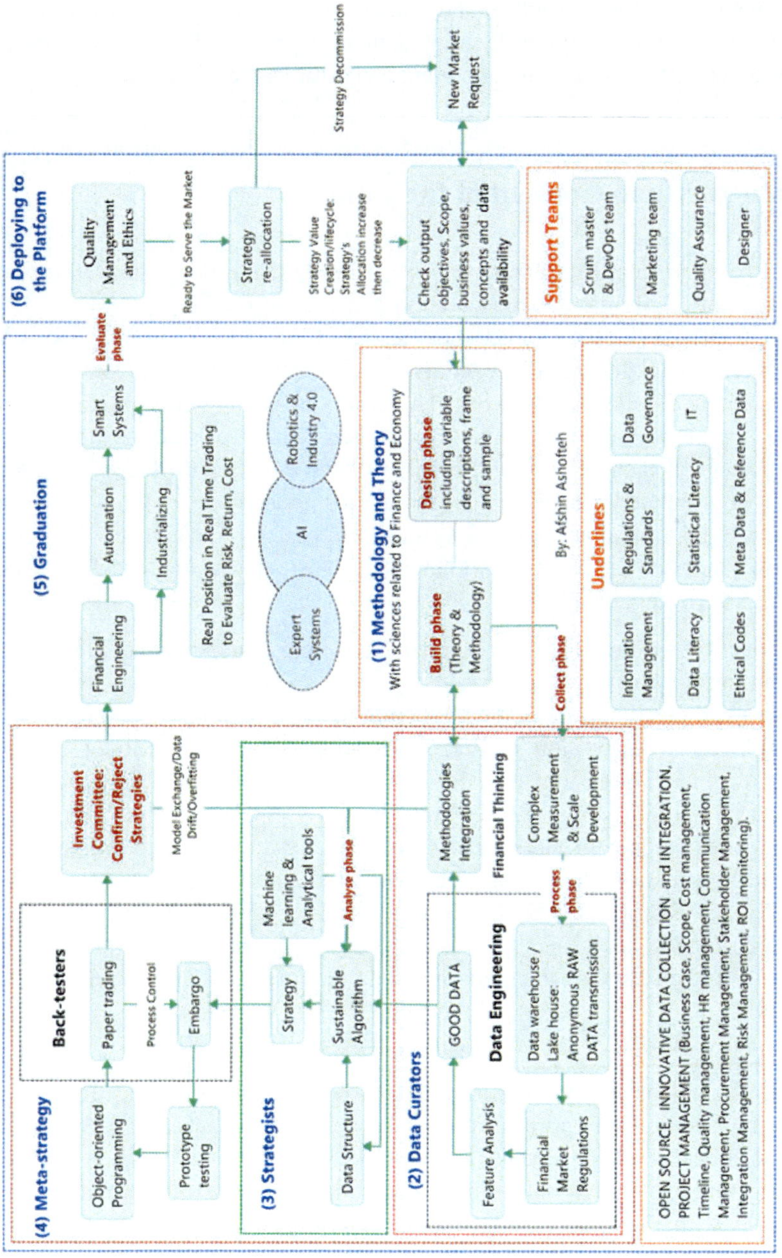

Fig. 1 Graphical summary of data science in Finance and its requirements

process in an electronic environment (see Fig. 2), which is a challenge, especially at the time of the COVID19 pandemic [10, 11].

Therefore, the remaining sections of the paper are organized as follows. In Sect. 2, we describe the structure and content of the proposed course in financial data science. Section 3 outlines the implementation process. The results are reported and discussed in Sect. 4. Finally, the main conclusions are presented in Sect. 5.

2 Structure and Content of the Course

The course is an interdisciplinary course. It needs theoretical knowledge, coding skills, and soft skills such as presentation and critical thinking [12]. Furthermore, it is an upper-level postgraduate course, enrolling mostly juniors and seniors with different backgrounds. The course needs a high level of communication, presentation, and collaborative work. Therefore, the desired educational objectives were set up; the contents were defined and organized, the proper teaching strategies were chosen for each section and topic, and the evaluation process to cover all activities was defined. Subsequently, the author revised the teaching strategies of the course to be a standalone course as much as possible. The course was organized as follows to offer the necessary knowledge of statistical and machine learning modeling for risk analysis:

1. Introduction to Financial data science, modeling concepts, and R/Python programming for finance.
2. Understand the importance and functioning of Regression for credit scoring (Simple linear regression model, Least squares criterion, Model evaluation, Multiple linear regression, Transformations, Model building, Regression pitfalls, Linear Probability Model (LPM), Logistic regression, Binary Probit model) [13], Time series models for market risk (Time-series patterns, Trend estimation, Seasonality estimation, smoothing methods, Stationarity, Autoregressive, Moving average, SARIMA, and Ensemble time series models) [14, 15], and Machine learning for risk analysis (supervised and unsupervised learning) [16].
3. Identify and distinguish the main modeling requirements and outcomes interpretation.
4. and finally, building useful reports for data-driven decisions.

The learning objective listed on the syllabus shows that learners would be capable of creating and implementing advanced modeling approaches to solve financial problems. Because of the diverse student body and different soft and hard skills, following learners individually and empowering them in these areas is almost impossible with the time limit of sessions and the traditional fixed design of classes. However, it would be possible if this interdisciplinary course is delivered by applying different means and technologies to answer different needs. To obtain this end, this course was preferred to be delivered by using different technologies, based on problem-based learning, active learning, Learning-by-doing, and hands-on approaches. It tries to empower learners in both the scientific part of modeling in finance and metacognitive and socioemotional skills in a constructive learning environment (see Fig. 1).

3 Implementation Process

The author's experience describes some suggestions for applying different technologies to the course presentation, intending to encourage relevant training in soft skills for persons involved in financial analysis and risk managers.[1] Considering all necessary elements was a big challenge, and the course was divided into three sections.

Part I of the course dealt with theoretical classes to involve students actively in the learning process. Preliminaries and slides provided an overview and information about the mechanics of the course. A forum on the Moodle provided an opportunity for participants to identify themselves and say a few words about their interest in the subject of the course. Students could refer to the shared information by their classmates to choose their team members for the projects and group activities. Part II with some support materials shared on the Moodle for self-study to adapt and learn more by themselves. In addition, the teacher provided supplementary handout materials in the format of articles, presentations, videos, and Q/As. Finally, part III to make an active contribution of students in defined activities as follows:

1. Defining some small tasks and an analytical modeling project in finance and asking learners to deliver the small tasks and the final project before deadlines distributed during the semester.
2. Making a discussion group with five critical questions extracted from the course's main concepts and asking learners to answer the questions and exchange ideas. It gives students this opportunity to see the comments of their colleagues and try to add more based on their knowledge, experience, and understanding of the theoretical classes.
3. Making a Kaggle competition in a teaching Kaggle profile and asking the learners to contribute to Kaggle's discussion about R/Python programming mainly related to the project of the course. This activity motivates students to analyze the data, practice data handling, and work on the issues raised on modeling and machine learning concepts or complex interactions among financial data, concepts, and programming.
4. Building a YouTube channel with short presentation videos of the instructor to demonstrate advanced concepts and essential takeaways related to the course and project as support materials. The videos were conducted with different levels and difficulties, from basic to advance levels.

[1] https://www.youtube.com/watch?v=Ajy9pOJ9DBc/.

5. Presenting chapters of the reference book: Marcos Lopez de Prado (2018), "Advances in Financial Machine Learning" by students and recording the presentations, each for a maximum of 20 min. Videos were shared for participants to watch at least three presentations of their colleagues, comment on divergent views on the topics according to their understanding, and raise the effectiveness of such a learning experience. It actively and extensively used the bibliography presented for this course and the chapters listed for each specific topic.
6. Sharing One-Pagers as a brief discussion about modeling concepts (only one page) and asking learners to read it and share their ideas in comments.

All these activities were supervised, reviewed, and evaluated by the lecturer. He answered the questions and corrected the wrong ideas. Each student's final grade consisted of 25% for answering the questions in the course discussion group and Kaggle, 25% for the book chapter presentation in a video format and commenting on other presentations, and 50% for the group project and problem-solving exercises. The deadlines for these activities were distributed during the semester.

In the broadest outline, the course and the underlying technology platforms are divided into six parts: Kaggle teaching facility,[2] YouTube channel of the course,[3] Discussion group,[4] R and Python programming platform in Kaggle,[5] Data and program sharing facilities of CODEOCEAN, and Zoom online sessions. All online materials may be copied and used for any non-commercial purpose.

Figure 2 represents the graphical summary of the main activities. Although the course was designed to serve financial problems with new scientific paradigms such as data science, the discussed approach will also be valuable and exciting if adapted to other traditional courses. For instance, the author has adopted the same approach for the Banking and Insurance Operations course.

4 Results

The course evaluation shows an average of 4.5/5, with all individual questions higher than 4. This combination of activities was accessible online for all students. It could provide an active learning atmosphere and motivate students to participate and share their knowledge, express themselves and try to communicate and solve the group project during the semester. Students' contributions on different platforms for different activities and learning from each other inspired them to participate actively in the learning process. Short articles provided by the teacher in the discussion group received 57 comments from students. Additionally, 250 comments from students on

[2] https://www.kaggle.com/c/credit-score-fall21/overview.

[3] https://www.youtube.com/channel/UCTOuxIhJxcxNOntTpamJeAA.

[4] https://www.linkedin.com/groups/12420006/.

[5] https://www.kaggle.com/aashofteh/.

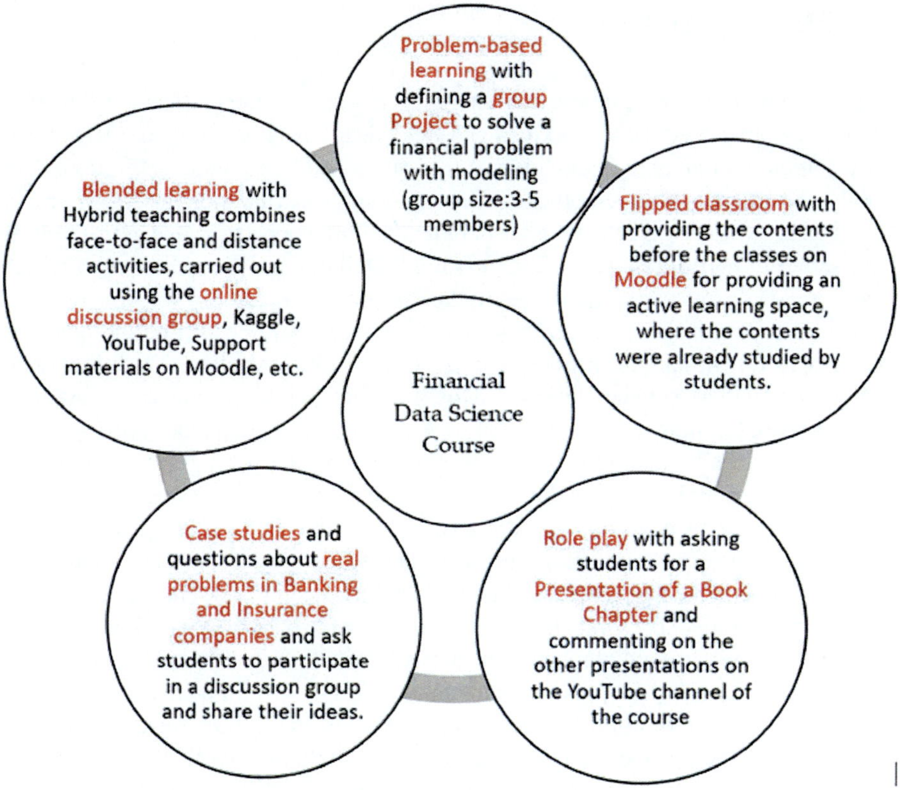

Fig. 2 Pedagogical innovation for Financial Data Science Course at the time of COVID-19 pandemic

a discussion about the course's concepts, 15 group presentations of the reference book chapters and 31 videos as support materials on YouTube, and 94 answers to questions about the programing project in Q/As on Kaggle that students tried to help each other, even if they were not in the same project group. International students had this opportunity to discuss their own country's local issues and possible solutions and share a copy of the indigenous market specifications, if it exists, along with the most recent edition of their national frameworks for risk management. For programming, we had 4 shared datasets in Kaggle and 10 R or Python shared codes to help students in their projects according to their questions. Sharing codes for everyone could make it fair for all groups to benefit from the teacher's help in coding. Finally, small group exercises were delivered to introduce some local and international issues related to risk analysis. These exercises have received attention with some relevant issues brought up by participants.

5 Main Conclusions

The proposed method's innovation integrates six free online platforms for teaching a course with a reasonable workload and exact deadlines distributed in one semester. This method replaces the evaluation of students based on different activities individually and as a group project. It considers this complex evaluation instead of only the final exam, which was interesting for students. Students participated actively during the semester in different parts, working individually on the presentations, commenting on videos, answering questions in the discussion group, and working as a team on their coding project. As all these activities were designed on online platforms, there was no limitation on time or place. The theoretical classes were conducted in person and as standard classes. This approach could stimulate a vibrant and constructive learning environment online. It had a pretty acceptable contribution rate of students. The high number of contributions in the discussion groups, commenting on videos, and high accuracy of models constructed by students on Kaggle highlight the importance of implementing these methodologies as a pedagogical innovation in higher education to facilitate the learning process by using new technologies besides in-person classes. According to this experience, there were some challenges and shortcomings. The first shortcoming of this approach was the time allotted to review and supervise all platforms and activities, including discussions, small group exercises, and responding to questions. In addition, it was time-consuming for only one presenter to monitor all these activities and evaluate them for the final grade. However, the active and friendly atmosphere of the course was the main driver of this enjoyable experience, and available technological facilities in the classrooms motivated students to learn and implement their knowledge in practice. Other major issues that the presenter needed to decide upon before offering this course were the number of participants with different backgrounds, and the adaptations of this multidisciplinary course to the experience and qualifications of the participants. As there is growing recognition of the importance of the data science application, even some professionals in risk management may participate in refresher training. As a result, the presenter should have not only strong teaching skills, leading discussions ability, and knowledge of analytical tools and programming skills [9], but also financial markets norms and regulations, with research interest related to data science and finance to be able to provide supplementary materials in elementary, intermediate, and advance levels. The last but not least challenge is accessing financial microdata from the financial institutions of interest. In the past decade, a policy revolution has taken place among financial authorities to recognize financial microdata as confidential personal information, which should not be disseminated along with conventional publications. As a result, providing updated, anonymized, and integrated financial microdata for each chapter of the course is challenging. Thanks to some public datasets from financial institutions, one large dataset for credit risk was built for this course and shared online with the public. Access to the microdata for credit scoring example is made available at https://codeocean.com/capsule/0503126/tree/v1 at no cost.

References

1. Ashofteh, A., Bravo, J.M.: A study on the quality of novel coronavirus (COVID-19) official datasets. Stat. J. IAOS **36**(2), 291–301 (2020). https://doi.org/10.3233/SJI-200674
2. Ashofteh, A.: Mining Big Data in statistical systems of the monetary financial institutions (MFIs). In: International Conference on Advanced Research Methods and Analytics (CARMA) (2018). https://doi.org/10.4995/carma2018.2018.8570
3. Longbing, C.: AI in finance: challenges, techniques, and opportunities. ACM Comput. Surv. **55**(3), 1–38 (2022). https://doi.org/10.1145/3502289
4. Perron, B.E., Victor, B.G., Hiltz, B.S., Ryan, J.: Teaching note—data science in the MSW curriculum: innovating training in statistics and research methods. J. Soc. Work Educ. 1–6 (2020). https://doi.org/10.1080/10437797.2020.1764891
5. Rizun, N., Nehrey, M., Volkova, N.: Data science in economics education: examples and opportunities, 550–564 (2022). https://doi.org/10.5220/0010926100003364
6. Saura, J.R., Ribeiro-Soriano, D., Palacios-Marqués, D.: Assessing behavioral data science privacy issues in government artificial intelligence deployment. Gov. Inf. Q. 101679 (2022). https://doi.org/10.1016/J.GIQ.2022.101679
7. Ashofteh, A., Bravo, J.M.: Data science training for official statistics: a new scientific paradigm of information and knowledge development in national statistical systems. Stat. J. IAOS **37**(3), 771–789 (2021). https://doi.org/10.3233/SJI-210841
8. Cahill, K., et al.: Building a Computational and Data Science Workforce. jocse.org (2022). https://doi.org/10.22369/issn.2153-4136/13/1/5
9. Bonnell, J., Ogihara, M., Yesha, Y.: Challenges and issues in data science education. Computer (Long. Beach. Calif). **55**(2), 63–66 (2022). https://doi.org/10.1109/MC.2021.3128734
10. Nacheva, R.: Emotions mining research framework: higher education in the pandemic context. Contrib. Econ., pp. 299–310 (2022). https://doi.org/10.1007/978-3-030-85254-2_18/COVER
11. Sakamaki, K., Taguri, M., Nishiuchi, H., Akimoto, Y., Koizumi, K.: Experience of distance education for project-based learning in data science. Jpn. J. Stat. Data Sci. 1–11 (2022). https://doi.org/10.1007/S42081-022-00154-2/TABLES/2
12. Donoho, D.: 50 years of data science. J. Comput. Graph. Stat. **26**(4), 745–766 (2017). https://doi.org/10.1080/10618600.2017.1384734
13. Ashofteh, A.: Big data for credit risk analysis: efficient machine learning models using PySpark. In: Proceedings of SIMSTAT 2019-10th International Workshop on Simulation and Statistics (2019)
14. Ashofteh, A., Bravo, J.M., Ayuso, M.: An ensemble learning strategy for panel time series forecasting of excess mortality during the COVID-19 pandemic. Appl. Soft Comput. **128**, 109422 (2022). https://doi.org/10.1016/j.asoc.2022.109422
15. Ashofteh, A., Bravo, J.M.: Life table forecasting in COVID-19 times: an ensemble learning approach. In: 16th Iberian Conference on Information Systems and Technologies (CISTI), pp. 1–6 (2021). https://doi.org/10.23919/CISTI52073.2021.9476583
16. Ashofteh, A., Bravo, J.M.: A conservative approach for online credit scoring. Expert Syst. Appl. **176**, 114835 (2021). https://doi.org/10.1016/j.eswa.2021.114835

Analysing Misclassifications in Confusion Matrices

Inmaculada Barranco-Chamorro and Rosa M. Carrillo-García

Abstract Techniques to deal with the off diagonal elements in confusion matrices are proposed. They are tailored to detect problems of bias of classification among classes. A Bayesian approach is developed aiming to estimate overprediction and underprediction probabilities among classes.

Keywords Confusion matrix · Misclassification · Overprediction · Underprediction

1 Introduction

Confusion matrices are the standard way of summarizing the performance of a classifier. It is assumed that the qualitative response to be predicted has $r \geq 2$ categories, the confusion matrix will be a $r \times r$ matrix, where the rows represent the actual or reference classes and the columns the predicted classes (or vice versa). So the diagonal elements correspond to the items properly classified, and the off-diagonal to the wrong ones. Most papers dealing with confusion matrices focus on the assessment of the overall accuracy of the classification process, such as kappa coefficient, and methods to improve these measurements, see for instance Liu et al. [8], Pontius and Millones [10], Congalton and Green [4], Grandini et al. [6] and references therein. Few papers consider the study of the off-diagonal cells in a confusion matrix. In this paper a method is proposed that can be useful for a better definition of classes and to improve the global process of classification.

Based on the results given in Barranco-Chamorro and Carrillo-García [2], first the problem of *classification bias* is introduced. This is a kind of systematic error, which may happen between categories in a specific direction. If a classifier is fair

I. Barranco-Chamorro (✉) · R. M. Carrillo-García
Department of Statistics and Operations Research, Faculty of Mathematics, University of Seville, Sevilla, Spain
e-mail: chamorro@us.es

© The Author(s), under exclusive license to Springer Nature Switzerland AG 2023
C. P. Kitsos et al. (eds.), *Statistical Modelling and Risk Analysis*,
Springer Proceedings in Mathematics & Statistics 430,
https://doi.org/10.1007/978-3-031-39864-3_3

or unbiased, then the errors of classification between two given categories A and B must happen randomly, that is, it is expected that they occur approximately with the same relative frequency in every direction. Quite often, this is not the case, and a kind of systematic error or bias occurs in a given direction. The classification bias can be due to deficiencies in the method of classification. For instance, it is well known, Goin [5], that an inappropriate choice of k in the k-nearest neighbor (k-nn) classifier may produce this effect. In case of being detected, the method of selection of k must be revised. On the other hand, the classification bias may be caused by the existence of a unidirectional confusion between two or more categories, that is, the classes under consideration are not well separated. Anyway, if this problem is detected, the process of classification should be improved. To identify this problem in a global way, first marginal homogeneity tests are proposed. The tests are based on the Stuart-Maxwell test, [3], and Bhapkar test, [12]. If the null hypothesis of marginal homogeneity is rejected, a *One versus All* methodology is proposed, in which McNemar type tests, [9], are applied to every pair of classes. Second a Bayesian method based on the Dirichlet-Multinomial distribution is developed to estimate the probabilities of confusion between the classes previously detected. Thus it can be assessed in a formal way, if certain classes suffer from a problem of overprediction or underprediction. To illustrate the use of our proposal, real applications are considered. As computational tools, we highlight that the R Software and R packages are used, [11].

2 Methodology

First we propose to apply techniques suited for *paired observations* to a confusion matrix. Let us introduce the appropriate notation.

Let Y and Z be two categorical variables with $r \geq 2$ categories. Y will be the variable that denotes *the reference (or actual) categories* and Z *the predicted classes*. As a result of the classification process, the confusion matrix given in Table 1 is obtained, where $n_{i,j}$ denotes the number of observations in the (i, j) cell for $i,\ j = 1, 2, \ldots, r$.

Table 1 Confusion matrix

	Z				
Y	1	2	$\cdots$	$r-1$	r
1	$n_{1,1}$	$n_{1,2}$	$\cdots$	$n_{1,r-1}$	$n_{1,r}$
2	$n_{2,1}$	$n_{2,2}$	$\cdots$	$n_{2,r-1}$	$n_{2,r}$
$\vdots$	$\vdots$	$\vdots$	$\vdots$	$\vdots$	$\vdots$
$r-1$	$n_{r-1,1}$	$n_{r-1,2}$	$\cdots$	$n_{r-1,r-1}$	$n_{r-1,r}$
r	$n_{r,1}$	$n_{r,2}$	$\cdots$	$n_{r,r-1}$	$n_{r,r}$

Usually global measurements are used to assess the performance of a classifier, such as Accuracy, Kappa index, Sensitivity, Specificity, Matthew's correlation coefficient, F1-score, and AUC. We highlight that all of them are global measurements, focusing mainly on the proportion of items properly classified, and they do not pay attention to structure in the off-diagonal elements.

3 Marginal Homogeneity

Next proper notation to deal with marginal homogeneity is introduced.

Notation The probability that (Y, Z) falls in the cell which corresponds to the ith row and the jth column is denoted as

$$\pi_{ij} = P[Y = i,\ Z = j].$$

$\{\pi_{ij}\}$ is the joint probability mass function (pmf) of (Y, Z).

- The marginal pmf of Y, denoted as $\{\pi_{i+}\}$ is

$$\pi_{i+} = \sum_{j=1}^{r} \pi_{ij} .$$

- The marginal pmf of Z, $\{\pi_{+j}\}$, is

$$\pi_{+j} = \sum_{i=1}^{r} \pi_{ij} .$$

$\{\pi_{i+}\}$ and $\{\pi_{+j}\}$ are the basis for constructing marginal homogeneity tests.

As already stated, cells in a confusion matrix are going to be handled as matched pairs of classes. We propose to test if marginal homogeneity can be assumed between the rows and the columns in this matrix, which is equivalent to testing if the row and column probabilities agree for all the categories, i.e.

$$P[Y = s] = P[Z = s] \quad \Longleftrightarrow \quad \pi_{s+} = \pi_{+s} \quad \forall s = 1, 2, \ldots, r . \tag{1}$$

Note that (1) states that the proportion of items classified in the sth class agrees with the proportion of actual or reference items in this class. If this agreement happens for all the categories, then this fact suggests that there do not exist systematic problems of classification (or classification bias) in our confusion matrix. This is the main idea on which to build this part of our proposal.

Method for a 2 × 2 Table McNemar type tests tailored for this context are proposed. For $i = 1, 2$, let us consider:

$$\begin{cases} H_0 : \pi_{i+} = \pi_{+i} \\ H_1 : \pi_{i+} \neq \pi_{+i} \; . \end{cases} \tag{2}$$

It can be seen in Barranco-Chamorro and Carrillo-García [2] that (2) is equivalent to:

$$\begin{cases} H_0 : \pi_{12} = \pi_{21} \\ H_1 : \pi_{12} \neq \pi_{21}. \end{cases} \tag{3}$$

In a classification problem, one of the variables refers to the actual category and the other one to the predicted class. So, in this context, the null hypothesis H_0 establishes that the probability of the class to be predicted is equal to the proportion of actual elements in the ith class. This agreement suggests that the performance of our classifier is good. Otherwise, the alternative hypothesis establishes that these probabilities significantly disagree, that is, there exists significant evidence of problems with the category under study. We highlight that this test allows us to focus on the probabilities associated with the off-diagonal elements in a confusion matrix, that is, the probabilities of the misclassified elements.

We recall that the McNemar test for 2×2 tables can be executed following a binomial approach, which allows us to carry out two-sided and one-sided tests, details can be seen in Barranco-Chamorro and Carrillo-García [2]. An asymptotic approach using chi-squared type statistics can also be considered. In our applications, we will follow the binomial approach, since this one allows us to carry out one-sided tests.

General Case, $r > 2$ In this setting, we have a confusion matrix resulting from a multi-class classifier with $r > 2$. The Stuart-Maxwell test (or Generalized McNemar test) can be applied, Sun and Yang [12]. The aim is to find evidence of significant differences between the actual and predicted probabilities in any of the categories, specifically, we test

$$\begin{cases} H_0 : \quad \pi_{i+} = \pi_{+i} \qquad \forall i = 1, 2, \ldots, r, \\ H_1 : \exists i \mid \pi_{i+} \neq \pi_{+i} \; . \end{cases} \tag{4}$$

The test is based on the vector of paired differences $\mathbf{d} = (d_1, \ldots, d_{r-1})$, where $d_s = \pi_{+s} - \pi_{s+}$.

Under the H_0 of marginal homogeneity, it was proven in Sun and Yang [12] that $E(\mathbf{d}) = 0$ and the asymptotic distribution of the following statistic can be approximated by a chi-square distribution with $(r-1)$ degrees of freedom

$$\chi_0^2 = N\mathbf{d}^t\widehat{\mathbf{V}}^{-1}\mathbf{d} = N\mathbf{d}^t(N\widehat{\mathbf{V}})^{-1}N\mathbf{d} \sim \chi_{r-1}^2, \tag{5}$$

where $N = \sum_{i,j} n_{i,j}$ and $\widehat{\mathbf{V}}$ is the estimated covariance matrix of vector $\sqrt{N}\mathbf{d}$, whose elements are given by

Table 2 Table 2 × 2

	$Z = i$	$Z \neq i$
$Y = i$	n_{ii}	$n_{i+} - n_{ii}$
$Y \neq i$	$n_{+i} - n_{ii}$	$\sum_{k \neq i} \sum_{j \neq i} n_{kj}$

$$\hat{v}_{st} = -(\pi_{st} - \pi_{ts}) \qquad s \neq t, \quad t, s = 1, \ldots, r-1,$$
$$\hat{v}_{ss} = \pi_{s+} + \pi_{+s} - 2\pi_{ss} \qquad t, s = 1, \ldots, r-1 .$$

A similar test was proposed by Bhapkar, details can be seen in Sun and Yang [12].

The null hypothesis, H_0, is rejected if and only if $p - value = P\left[\chi^2_{r-1} \geq \chi^2_{obs}\right]$ is less than the fixed significance level of test, α, where χ^2_{obs} denotes the observed value of applying χ^2_0 statistic to our matrix.

That is, for $r > 2$, we propose to test multiple marginal homogeneity by using Stuart-Maxwell or Bhapkar tests. If the null hypothesis of marginal homogeneity is rejected, the next step is to look for those categories with serious deficiencies in the classification problem, that is to carry out *post hoc tests* to explore which categories are significantly different while controlling the experiment-wise error rate. Bonferroni corrections are considered in our applications. McNemar tests for 2×2 tables will be proposed. So, for the ith category, with $i = 1, \ldots, r$, let us consider Table 2 obtained from Table 1, and we test

$$\begin{cases} H_{0,i} : P[Y = i, Z \neq i] = P[Y \neq i, Z = i] \\ H_{1,i} : P[Y = i, Z \neq i] \neq P[Y \neq i, Z = i]. \end{cases} \tag{6}$$

$H_{0,i}$ states that the proportion of elements, which belong to the ith class ($Y = i$) but are classified into other ones ($Z \neq i$) must agree with the proportion of elements which belong to the remaining classes ($Y \neq i$) and have been misclassified in the ith category ($Z = i$). The McNemar test, can be applied to every table with the statistic test T_i, which follows, under $H_{0,i}$, a binomial distribution with success probability 0.5

$$T_i = n_{i+} - n_{ii} \sim_{H_0} B(n_{i+} + n_{+i} - 2n_{ii},\ 0.5).$$

Details about the use of statistics T_i are given in Sect. 5.

4 Bayesian Approach

Once the classes with problems are detected a Bayesian methodology is proposed to estimate the probabilities of misclassification. Our approach is based on the *multinomial-Dirichlet distribution* appropriate for a confusion matrix. For the matrix introduced in Table 1, note that the number of elements in the kth row, denoted as n_{k+}, is fixed (since the rows are the actual or reference categories). Our proposal is to deal with the kth row, $\mathbf{Y}_k$, as a multinomial distribution with n_{k+}

trials and r possible outcomes (these are to be classified in the $\{1, \ldots, r\}$ classes). The elements of $\mathbf{Y}_k$ are denoted as $Y_{j|k}$, where $Y_{j|k}$ counts the number of elements in the kth reference category classified in the jth class, for $j = 1, \ldots, r$, $\mathbf{Y}_k = (Y_{1|k}, \ldots, Y_{r|k})$. The corresponding probabilities are denoted as $(\theta_{1|k}, \ldots, \theta_{r|k})$. That is

$$(\mathbf{Y}_k | n_{k+}, \boldsymbol{\theta}_k) \sim Multinomial(n_{k+}, \boldsymbol{\theta}_k) \quad \text{where} \quad \boldsymbol{\theta}_k = (\theta_{1|k}, \ldots, \theta_{r|k}),$$

Remark We highlight that, in terms of the previously introduced notation,

$$\theta_{j|k} = P[Z = j | Y = k].$$

As prior distribution for $\boldsymbol{\theta}_k$ a *Dirichlet distribution* is proposed

$$(\boldsymbol{\theta}_k | \boldsymbol{\alpha}_k) \sim Dirichlet(\boldsymbol{\alpha}_k) \ ,$$

where $\boldsymbol{\alpha}_k = (\alpha_{1|k}, \ldots, \alpha_{r|k})$ with $\alpha_{j|k} \geq 0$.

Conjugacy in the Dirichlet-Multinomial Given a confusion matrix, whose observed rows are denoted by

$$\mathbf{y}_k^{obs} = (n_{k,1}, \ldots, n_{k,r}) = (n_{1|k}, \ldots, n_{r|k}),$$

by applying Bayes Theorem, and since the Dirichlet distribution is a conjugate prior for the Multinomial model, the posterior distribution for $\boldsymbol{\theta_k}$ is

$$\pi(\boldsymbol{\theta_k} | \mathbf{y}_k^{obs}, \boldsymbol{\alpha_k}) \propto \prod_{j=1}^{r} \theta_{j|k}^{n_{j|k} + \alpha_{j|k} - 1} \ ,$$

where $\propto$ stands for proportional to.

Therefore,

$$\boldsymbol{\theta_k} | \mathbf{y}_k^{obs}, \boldsymbol{\alpha_k} \sim Dirichlet(n_{1|k} + \alpha_{1|k}, \ldots, n_{r|k} + \alpha_{r|k}) \ . \tag{7}$$

Bayesian inference can be carried out. Main advantages of (7) are:

(i) The mode and mean of (7) have closed expressions.
(ii) The *marginals of (7)* are Beta distributed, and therefore manageable.
(iii) Credible intervals for the parameters can be given, which also provide a measurement of uncertainty.

5 Applications

5.1 Application 1

A confusion matrix taken from the fields of Geostatistics and Image Processing, given in Congalton and Green [4] is considered, (Table 3). The matrix has four categories ($r = 4$) and was obtained from an unsupervised classification method from a Landsat Thematic Mapper image. The categories related to the land use are: *FallenLeaf*, *Conifers*, *Agricultural* and *Scrub*. Rows correspond to the Actual classes and columns to the Predicted classes. The sample size is $n = 434$. As for a global measurement of classification, we have that the $accuracy = 0.74$.

The multiple marginal homogeneity Stuart-Maxwell and Bhapkar tests were applied in Barranco-Chamorro and Carrillo-García [2]. We reached the conclusion that there exists significant evidence *to reject the null hypothesis of marginal homogeneity* for $\alpha = 0.05$. The next step is *to look for those categories with deficiencies* in the classification process. One-sided and two-sided McNemar tests are applied for every category by considering the subtables resulting from applying Table 2. As an example, let us consider the Fallen Leaf class, whose subtable is given in Table 4. The subscript *fl* will be used to emphasize that we refer to FallenLeaf class. We propose to testing

$$\begin{cases} H_{0,fl} : P[A\_FallenLeaf \cap P\_Others] \geq P[A\_Others \cap P\_FallenLeaf] \\ H_{1,fl} : P[A\_FallenLeaf \cap P\_Others] < P[A\_Others \cap P\_FallenLeaf]. \end{cases} \tag{8}$$

(8) is referred as **Less** in Table 5. Following the binomial approach, (8) is equivalent to testing that $p_{fl} = P[A\_FallenLeaf \cap P\_Others]$ is *less than* 0.5

$$\begin{cases} H_{0,fl} : p_{fl} \geq 0.5 \\ H_{1,fl} : p_{fl} < 0.5. \end{cases} \tag{9}$$

Table 3 Confusion matrix: land use

	P_FallenLeaf	P_Conifers	P_Agricultural	P_Scrub
A_FallenLeaf	65	6	0	4
A_Conifers	4	81	11	7
A_Agricultural	22	5	85	3
A_Scrub	24	8	19	90

Table 4 Auxiliary table for **FallenLeaf**

	P_FallenLeaf	P_Others
A_FallenLeaf	65	10
A_Others	50	309

Table 5 McNemar test for every category

	FallenLeaf	Conifiers	Agricultural	Scrub
Less	1×10^{-7}	0.7336454	0.5512891	0.9999994
Greater	1.0000	0.3776143	0.5512891	0.0000022
Two_Sided	2×10^{-7}	0.7552287	1.0000000	0.0000045

The p-value of (9) is $p\text{-}value = P[B(60, 0.5) \leq 10] = 1 \times 10^{-7}$ and therefore, we may conclude that there exist evidence of $p_{fl} < 0.5$, or equivalently, $P[A\_FallenLeaf \cap P\_Others] < P[A\_Others \cap P\_FallenLeaf]$.

By proceeding similarly, the one-sided and two-sided tests for every category, along with the p-values that we obtained, are given in Table 5. Significant p-values for the alternative hypothesis *Less* or *Greater* suggest the existence of an overprediction or underprediction problem, respectively, in the category under consideration, additional details can be seen in Barranco-Chamorro and Carrillo-García [2].

There exist evidence to reject the marginal homogeneity in the one-sided tests for the categories *FallenLeaf* and *Scrub*, which correspond to p-values 1×10^{-7} and 2.2×10^{-6} in Table 5, respectively. Therefore we may conclude that:

- For the *Fallen Leaf class:* There exist confusion between the rest of categories and $FallenLeaf$, since many more observations are assigned to *FallenLeaf* class than those that really belong to it. It could be said that there exists an *overprediction* problem of observations in the class $FallenLeaf$.
- For the *Scrub class:* An important part of them are predicted in other classes, therefore causing an *underprediction* misclassification problem.

Let us now estimate the probabilities of misclassification in these classes by using a Bayesian methodology. Credible intervals are also given.

Posterior Probabilities Estimates

- For every category, a uniform prior distribution is considered, which corresponds to the Dirichlet distribution with $\boldsymbol{\alpha_k} = (1, \ldots, 1)$. This can be considered as a non-informative prior.
- Table 6 is given with the Bayesian estimates of probabilities in every conditional distribution (by columns). These estimates were obtained as the mean of the posterior distribution.

As example, for reading Table 6, let us look at the fourth column in Table 6, where the conditional probabilities associated with $Actual\_Scrub$ category have been estimated. In this column, we have that

$$\widehat{P}[P\_Scrub|A\_Scrub] = 0.628$$

is quite low.

Table 6 Summary Bayesian estimates of conditional probabilities in the **Land use** problem

	A_FallenLeaf	A_Conifiers	A_Agricultural	A_Scrub
P_FallenLeaf	**0.835**	0.047	**0.193**	**0.172**
P_Conifers	0.089	**0.766**	0.05	0.062
P_Agricultural	0.013	**0.112**	**0.723**	**0.138**
P_Scrub	0.063	0.075	0.034	**0.628**

Moreover, we have that

$$\widehat{P}[P\_FallenLeaf|A\_Scrub] = 0.172$$

$$\text{and} \quad \widehat{P}[P\_Agricultural|A\_Scrub] = 0.138\ .$$

It could be said that there exists an underprediction of the Scrub category, since observations which are actual Scrub are often misclassified as FallenLeaf or Agricultural. The probabilities we have commented have been highlighted in bold in fourth column of Table 6. Also note that these appreciations are coherent with the result in Table 5.

As for the first column, corresponding to the conditional probabilities in the class $A\_FallenLeaf$, we highlight the good estimates obtained for

$$\widehat{P}[P\_FallenLeaf|A\_FallenLeaf] = 0.83\ ,$$

which is given in bold in the first column of Table 6. Similarly we have highlighted in bold the probabilities of interest in the rest of columns in Table 6.

However, note that, in the first row of Table 6, we have

$$\widehat{P}[P\_FallenLeaf|A\_Agricultural] = 0.19$$

$$\text{and} \quad \widehat{P}[P\_FallenLeaf|A\_Scrub] = 0.17,$$

which are also coherent with results in Table 5.

It could be said that those elements which are in the actual FallenLeaf class are properly classified, but there exists problems of confusion of other categories to Fallen Leaf, specifically actual Agricultural and Scrub observations are often misclassified as Fallen Leaf. Both facts cause an overprediction of the Fallen Leaf class.

5.2 *Application 2*

The confusion matrix is given in Table 7. This matrix is obtained as result of applying an artificial intelligence classification method for the diagnosis of Inflammatory Bowel Disease (IBD) based on fecal multiomics data [7]. IBD's are:

Table 7 Confusion matrix in Inflammatory Bowel Disease (IBD)

	P_nonIBD	P_UC	P_CD
A_nonIBD	37	1	15
A_UC	6	19	26
A_CD	15	3	77

- Crohn's disease (CD)
- Ulcerative Colitis (UC),

whereas nonIBD class refers to the control group.

These diseases are difficult to diagnose and classify, and their accurate diagnosis is really an important issue in Medicine.

Huang et al. [7] proposed a method with high accuracy. Specifically, the accuracy of Table 7 is $accuracy = 0.6683$. However certain asymmetry is observed in the off-diagonal elements of Table 7, which deserves additional analysis.

In this case, we are just going to list conclusions, details can be seen in Barranco-Chamorro and Carrillo-García [2]. The marginal homogeneity was again rejected. For the control group (nonIBD), we did not detect any systematic error.

- For the UC class, the one-sided McNemar test suggested underprediction.
- For the CD class, the McNemar test suggested overprediction.

Bayesian approach based on the Dirichlet multinomial with a noninformative prior was applied. We have $r = 3$ categories. As novelty, we highlight that the Dirichlet posterior distribution for every category can be represented in the two-dimensional simplex. So, in this case we have a *visual representation* of these posterior distributions.

The plot of the Dirichlet Posterior distribution in Control Group (nonIBD) is

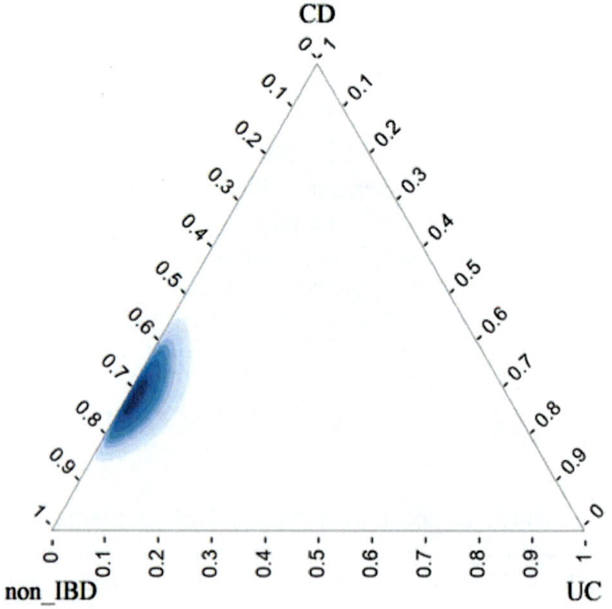

The associated Bayes estimates and 95% Credible Intervals are:

- $\widehat{P}[P\_nonIBD|A\_nonIBD] = 0.68$ and (0.55, 0.79).
- $\widehat{P}[P\_CD|A\_nonIBD] = 0.28$ and (0.18, 0.41).

Comments on these summaries:

- Although in the control group, non_IBD, there is no evidence of classification bias, *the estimated probability of being classified as CD is relatively high.*
- The joint posterior distribution is quite concentrated and close to non_IBD vertex.

Plot of the Dirichlet Posterior distribution in UC (Ulcerative Colitis)

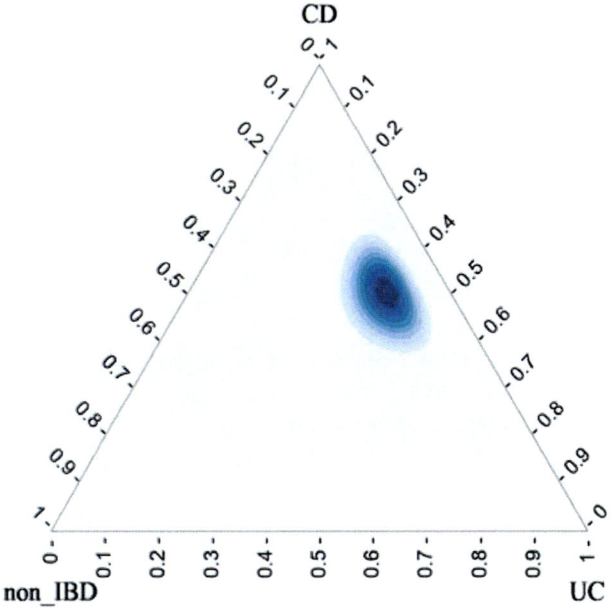

The Bayes estimates and 95% Credible Intervals are:

- $\widehat{P}[P\_UC|A\_UC] = 0.37$ (quite low) and (0.25, 0.50).
- $\widehat{P}[P\_CD|A\_UC] = 0.51$. Note that this probability is surprisingly high. The interval is (0.37, 0.63).

Comment on these summaries: The class *Ulcerative Colitis* suffers from *under-prediction.*

Plot of the Dirichlet Posterior distribution in CD (Crohn's Disease).

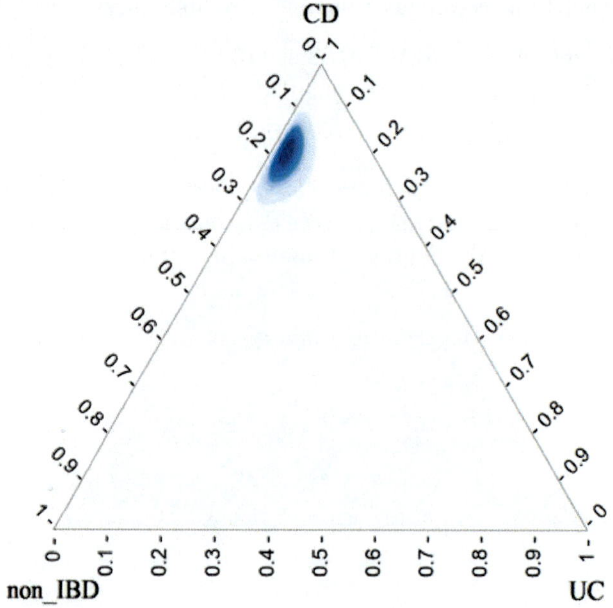

– Bayes estimates and 95% Credible Intervals in this class are

- $\widehat{P}[P\_CD|A\_CD] = 0.80$. Note that it is the highest one we get in this example. The credible interval is (0.71, 0.87)
- $\widehat{P}[P\_nonIBD|A\_CD] = 0.16$ and (0.10, 0.24).
- $\widehat{P}[P\_UC|A\_CD] = 0.04$ and (0.01, 0.09).

Comments on these summaries:

– The area of highest posterior density is close to CD vertex.
– The class *Crohn's Disease* suffers from *overprediction*.

6 Conclusions and Final Comments

In this paper, methods to detect the *bias of classification*, as well as *overprediction and underprediction problems* associated with categories in a confusion matrix, are proposed. They may be applied to results of applying supervised learning algorithms, such as logistic regression, linear and quadratic discriminant analysis, naive Bayes, k-nearest neighbors, classification trees, random forests, boosting or support vector machines, among others. Marginal homogeneity tests for matched pairs of observations are proposed.

Also the Multinomial-Dirichlet distribution can be applied to asses the probabilities of over- and under-prediction in a misclassification problem. Two applications taken from different scientific areas have been carried out. The results are satisfactory. Applications to efficient intrusion detection systems can be seen in Aldallal [1], and to agriculture in Wei et al. [13].

We highlight that our proposal is of interest for a better definition of classes, and to improve the performance of classification methods.

Acknowledgments The authors want to thank the referees for the careful reading of the paper and their helpful comments, which helped to improve it.

References

1. Aldallal, A.: Toward efficient intrusion detection system using hybrid deep learning approach. Symmetry **14**(9), 1916 (2022). https://doi.org/10.3390/sym14091916
2. Barranco-Chamorro, I., Carrillo-García, R.M.: Techniques to deal with off-diagonal elements in confusion matrices. Mathematics **9**(24), 3233 (2021). https://doi.org/10.3390/math9243233
3. Black, S., Gonen, M.: A generalization of the Stuart-Maxwell test. In: SAS Conference Proceedings: South-Central SAS Users Group. Applied Logic Associates, Inc., Houston (1997)
4. Congalton, R.G., Green, K.: Assessing the Accuracy of Remotely Sensed Data. Principles and Practices, 3rd edn. CRC Press, Boca Raton (2020)
5. Goin, J.E.: Classification bias of the k-nearest neighbor algorithm. IEEE Trans. Pattern Anal. Mach. Intell. **PAMI-6**, 379–381 (1984). https://doi.org/10.1109/TPAMI.1984.4767533
6. Grandini, M., Bagli, E., Visani, G.: Metrics for multi-class classification: an overview. arXiv (2020). arXiv:2008.05756
7. Huang, Q., Zhang, X., Hu, Z.: Application of artificial intelligence modeling technology based on multi-omics in noninvasive diagnosis of inflammatory bowel disease. J. Inflamm. Res. **14**, 1933–1943 (2021)
8. Liu, C., Frazier, P., Kumar, L.: Comparative assessment of the measures of thematic classification accuracy. Remote Sens. Environ. **107**, 606–616 (2007)
9. McNemar, Q.: Note on the sampling error of the difference between correlated proportions or percentages. Psychometrika **12**, 153–157 (1947)
10. Pontius, R., Millones, M.: Death to Kappa: birth of quantity disagreement and allocation disagreement for accuracy assessment. Int. J. Remote Sens. **32**, 4407–4429 (2011)
11. R Core Team: R: A language and environment for statistical computing. R Foundation for Statistical Computing, Vienna (2022) https://www.R-project.org/.
12. Sun, X., Yang, Z.: Generalized McNemar's test for homogeneity of the marginal distributions. In: Proceedings of the SAS Global Forum Proceedings. Statistics and Data Analysis, San Antonio, 16–19 March, vol. 382, pp. 1–10 (2008)
13. Wei, H., Chen, W., Zhu, L., Chu, X., Liu, H., Mu, Y., Ma, Z.: Improved lightweight mango sorting model based on visualization. Agriculture **12**(9), 1467 (2022). https://doi.org/10.3390/agriculture12091467

Management Excellence Model Use: Brazilian Electricity Distributors Case

Alexandre Carrasco, Marina A. P. Andrade, Álvaro Rosa, and Maria Filomena Teodoro

Abstract The article evaluates the impact of the use of the excellence management model (EMM) in Brasil by electricity distribution companies and their impact on customer satisfaction, that is, on the index of consumer satisfaction evaluation (ICSE). A total of 10 year were evaluated for the use of the model in groups of companies with different levels of model implantation (users, indifferent, engaged and winning) using statistical methods. As result, it was verified the existence of differences between the groups revealed the correct decision by the use of the model seen by this view. The results can be used by similar organizations or other industries.

Keywords Excellence model · Costumer satisfaction · Electricity distributors · Parametric and nonparametric approach

1 Introduction

Over the last few years, the Brazilian energy sector has been undergoing constant transformations, provoked by dynamic scenarios, a more restrictive regulatory environment, and specifically performed by industry associations like the Association of

---

A. Carrasco · Á. Rosa
ISCTE-IUL, Instituto Universitário de Lisboa, Lisboa, Portugal

M. A. P. Andrade (✉)
ISCTE-IUL, Instituto Universitário de Lisboa, Lisboa, Portugal

ISTAR, Information Sciences, Technologies and Architecture Research Center, Lisbon, Portugal
e-mail: marina.andraded@iscte-iul.pt

M. F. Teodoro
CEMAT, Center for Computational and Stochastic Mathematics, Instituto Superior Técnico, Lisbon University, Lisbon, Portugal

CINAV, Center of Naval Research, Naval Academy, Portuguese Navy, Almada, Portugal

© The Author(s), under exclusive license to Springer Nature Switzerland AG 2023
C. P. Kitsos et al. (eds.), *Statistical Modelling and Risk Analysis*,
Springer Proceedings in Mathematics & Statistics 430,
https://doi.org/10.1007/978-3-031-39864-3_4

Electric Power Distribution Companies (AEDC). The progressive improvement in the quality of services rendered, competitive prices, efficiency and cost-effectiveness [1], had been pushed by industry associations like AEDC, or the regulatory agencies who propose actions that enable the companies to face these challenges.

Among the actions implemented are the Index of Consumer Satisfaction Evaluation Index (ICSE) and the AEDC awards, promoted by Brazil's regulatory agency National Agency of Electric Energy (NAEE) and the (AEDC). In the case of the ICSE prize, NAEE annually conducts a satisfaction survey on services provided to residential consumers, which enables the ICSE prize to cover the entire national territory. The AEDC prize, more specifically in the Management Quality category, serves as an incentive to the adoption of the EMM (Excellence in Brazilian Management Model) provided by distribution companies. In this category, participation in the NQP (National Quality Prize) awards a grade that is subsequently used in the assessment of the AEDC Quality in Management Award.

The energy sector has participated significantly in the National Quality Prize (NQP) and in similar awards, both regional and sectoral. For Boulter et al. [2] and Corredor and Gofii [3] the organizations that obtain these awards in recognition for their accomplishments are those that have the best results. Thus, the objective is consistent with the primary aim of the award: to encourage improvements that yields results.

Here the period of 2007 to 2016 is where the focus is. The interest is to evaluate if there are differences in performance of the ICSE among the companies that regularly use the model, including the award-winning companies and the indifferent ones. Thus, this research is motivated by studying distributors and their impact on consumer satisfaction which may present two different perspectives. The user view, focusing on the impacts of lack of infrastructure as well as the inability to invest and the impacts of poor supply quality on the economy and population. And, the management view, where there are still difficulties to obtain studies that support decision to adopt models of excellence because there are different and antagonist studies, reinforcing the necessity of access and analyze restricted information in order to support the decision makers. This last perspective is the main interest in this work.

2 Literature Review

Customer satisfaction ratings can be found in relation to a product or process. It can be defined as a result of a consumption experience or as a consumer's response to a balance between expectations for a product or service and results obtained [4–6].

NAEE was used as an external benchmark and comparative tool, to measure the quality of service provided by assignee's of electrical energy utility services, and to address issues and improve regulations, study published by Marchetti and Prado [7, 8]. Since 2000, NAEE has been promoting the ICSE Awards, which recognize organizations with the highest scores in the ICSE Customer Satisfaction Survey [9].

Currently, all distribution companies are required to conduct surveys and are held accountable for their results, which are available for public consultation on the institution's website. Therefore, surveys serve as a mechanism of popular pressure used by regulators to encourage improvement of the services provided.

According to Oliver [10, 11], satisfaction can be understood as the evaluation of (user's) surprise in the consumption experience. Therefore, it is a relevant management tool for organizations looking to improve their service delivery.

In the case of public services such as electricity supply, this type of evaluation serves to enhance the process of monitoring the results of distribution companies and to revise and direct public policy, and straight sectoral efforts to meet consumer needs, Marchetti and Prado, 2004, (see [7]). Designed in this way for the energy distribution sector, the ICSE represents an index of high social and managerial relevance.

However, the national results show no significant improvement (in recent years). It indicates that there is sufficient scope to achieve a higher level of improvement compared to the comparative benchmark.

3 The Excellence Management Model

According to the National Quality Foundation (NQF) (2016), the excellence management model (EMM) is a world-class business management system or model for management excellence. For Puay et al. [12] and Miguel et al. [13], these models represent a country's efforts to enhance its international reputation in the global marketplace.

In the Brazilian context, the EMM model deserves special attention as it has become one of the most important guidelines for Brazil's competitiveness [14]. Its importance also confirms AEDC's efforts to promote activities that bring the best results to electrical energy distribution companies and areas of interest, such as the AEDC Award.

The universality of the model, its potential to be used by all types of organizations, see Calvo-Mora et al. [15], and slight differences in scope, see Bucelli and Costa [5] also contribute to the exchange of practices across industries, maximizing the benefits of its use.

Exchanges of knowledge may also occur in relation to practices adopted and/or results obtained from similar awards such as the Deming Prize (Japan), the Malcolm Baldrige National Quality Award (USA) and the European Quality Award (Europe). Khoo and Tan [16], Tan [17], and Puay et al. [12].

In this study, it is appropriate to emphasize that the version of the EMM model adopted between 2007 and 2016 consisted of two evaluation groups: process and outcome. Both are tuned to the basic principles of excellence.

Process-related criteria are established through the analysis of detailed information about how an organization implements its management processes without a predefined methodology. Information on leadership, strategy and planning, clients, social responsibility, information and knowledge, people and business processes is requested by each organization. In addition to benchmarking studies demonstrating evidence of leadership [18], the outcome item requires results demonstrating implementation of practices, achievement of strategic outcomes, and fulfillment of stakeholder-supplied requirements over a period of at least three years.

Assessments are derived from a panel of judges formed for this purpose, comprised of raters who are experts in various fields of knowledge, training, and backgrounds. As a result, evaluated organizations receive a score ranging from 0 to 1000 based on pre-set criteria and written comments on the strengths and areas for improvement that can be used to improve management. However, the benefits of their use are still controversial and require further analysis individually for each sector or performance attribute. For Doelema et al. [19], despite its widespread use, the implementation success of this model is not guaranteed. For Boulter et al. [2] and Corredor and Goñi [3], organizations that adopt the reference model show superior results.

In the context of this study, Puay et al. [12] points out that the users of excellence models not only improve their quality but also enhance other performance attributes, including client satisfaction, which is represented in the electric energy sector in the form of the ICSE. Thus, the relation between GS (Global Score) and the ICSE, and the difference in ICSE performance among different groups of companies involved (users vs. indifferent, award winning vs. engaged) will be our object of study. The period in analysis is 2007 up to 2016 and the objectives were divided in two different hypotheses, the hypotheses, which have been substantiated in the course of the main analysis, are : $HA1$—User organizations deliver a higher level of performance in ICSE than indifferent organizations, and, secondly, $HB1$—Award winning organizations deliver a higher level of performance in ICSE than engaged organizations.

4 Methodology

4.1 Research Classification—Sample and Population

This research study was classified according to a proposition put forward by Vergara [20], which qualifies it in relation to ends and means. As regards ends, this is an exploratory, descriptive and explanatory research. As for means, the research is bibliographical, documental and ex post facto.

The sample is composed of 31 organizations that correspond to 96.1% of the total number of consumers and 95% of the electric energy distributed nationally. The companies selected contain public data made available through sustainability reports, which imparts a necessary level of maturity to the process of collating the information, which enables it to be made publicly available while ensuring an adequate degree of transparency.

The period of analysis comprehends the years between 2007 and 2016, where the initial year of the series relates to that when the NQP started to be used as criteria for awarding the AEDC Management Quality Prize, in a correlated way. The organizations participating in the study are listed in the table (table was obtained from [6], Carrasco master thesis author's production).

4.2 Data Collection and Variables of Interest

In this study the Global Score (GS) and ICSE variables were the analyzed variables. The GS variable is derived from the scores earned by the NQP's companies and ranges from 0 to 1000 points, which represents the level of maturity of management practice. The collection process for this variable reflects the judgment of a multidisciplinary committee with training and experience in evaluating good models. Additionally, the data used to categorize organizations according to participation was provided by the NQF, where award winners were made public through press releases and market announcements and posted on corporate websites.

The ICSE framework, in turn, is organized annually by the NAEE using standard methodologies, Marchetti and Prado [7, 8]. Its value varies from 0 to 100% and is available for public consultation on the corporate website from the first year the award application is submitted. The data sets provided are identical to those used for the ICSE award (NAEE 2016).

Information is organized and tabulated in a way that is analyzed and distilled. In all cases, organizations have enough data to perform their intended analysis.

Organizations are divided into two groups, a group in which are considered the users and the indifference organizations. Afterwards, the participating organizations were subdivided into another group with the winners and the engaged organizations. Therefore, it may be summarized as:

users—that is, the organizations that have over 3 participations or attained 30% in the period of analysis, which characterizes regular participation;

prize winners—that is, the user organizations that obtained greater recognition in the period of analysis;

engaged—that is, the user organizations that did not receive awards;

indifferent—that is, those organizations that participated up to 3 times or 30% of the number of times during the period analyzed, sequentially or not.

Given the formed groups, a descriptive statistics analysis was performed considering the collected variables, subsequently the same analysis was considered for the grouped data. The central location and dispersion measures were obtained along with a graphical analysis – histogram, dispersion graphs and boxplot.

The choice of tests used the Anderson-Darling test, [21] to measure the assumption of normality required for some types of analysis. The tests used showed that the ICSE variable was not normally distributed at the 5% confidence level. As suggested by Pino [21], the chosen option for dealing with nonnormality was to use nonparametric tests. The Kruskal-Wallis test was deemed suitable for this purpose and was applied. All tests and analyzes were performed using *Minitab* 10.0 and *Excel* software. In summary, the GS data comes from an assessment made by NQF by trained auditors in accordance with EMM while ICSE comes from a survey carried out by NAEE with the distributor's customers.

5 Data Analysis

A descriptive statistics data analysis was performed and the database contents were individually analyzed and corrected as necessary, paying particular attention to missing data and unstructured data (outliers). In Table 1 are presented some important descriptive statistics of ICSE and GS for the groups of interest in this study. With respect to GS, missing data were supplemented by the results of the last participation of the NQP. When organizations cease to participate, naturally practices are interrupted, given the opportunity to occur gaps. For ICSE, 2011 data have not been verified by NAEE, and NAEE has not distributed the indices obtained, Carrasco [6]. Hence we opted for repeating the values of 2010 for the period of 2011. No correction of the atypical data (outliers) was necessary.

Figures 1 and 2 present the box–plot of each variable. We observe some similarities in the representations, concerning the positioning of each box with respect to the assigned group—users, winners, engaged and indifferent. When evaluating the measures of position, it can be observed that the best ICSE, as well as, GS results can be found in the group of award winner companies, whereas the worst relate to the indifferent ones. The dispersion measures also corroborate this analysis. Even though the correlation between the indexes is low, it can be noted that the award winner organizations exhibit higher of the indexes than the engaged, or the indifferent companies. The results of the dispersion measures reinforce the low correlation, between the two indexes, established in the Pearson correlation test (0.18803). It also occurs due to the fact that the ICSE represents the satisfaction of consumers, therefore providing an external view of the organization, while the GS

Table 1 Some descriptive measures for GS and ICSE *per* group

	Variable	Min.	Mean	Median	Max.
Indifferent	GS	99.0	212.6	176.3	376.3
	ICSE	35.5	55.8	56.3	71.8
Users	GS	163.0	515.8	512.4	800.0
	ICSE	48.1	64.3	64.4	79.0
Engaged	GS	163.0	444.3	429.3	683.0
	ICSE	48.1	63.1	63.0	77.9
Winners	GS	409.0	634.4	667.0	800.0
	ICSE	51.7	66.5	67.8	79

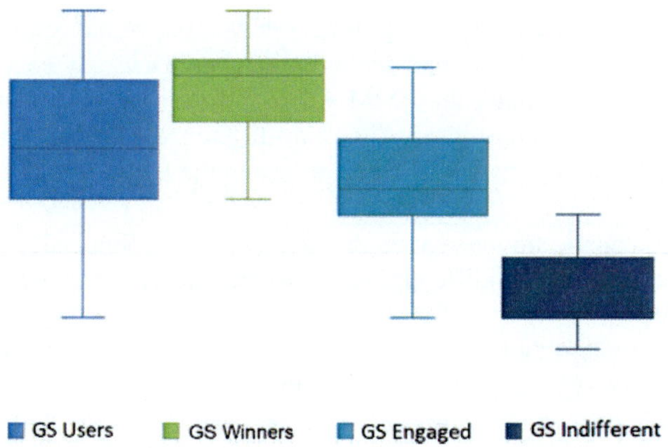

Fig. 1 GS box-plot *per* group

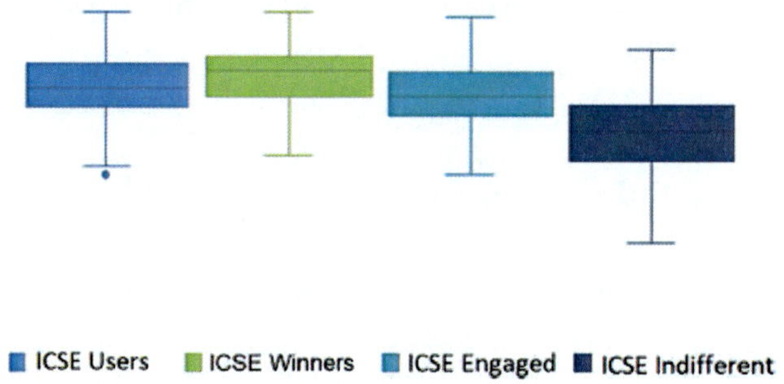

Fig. 2 ICSE box-plot *per* group

indicates internal improvements, as well as those related to the management of the indicators that will reflect the highest satisfaction indexes. Therefore, ICSE an GS, may be considered as complementary.

For each hypothesis being tested, we used the Kruskal-Wallis test [22], to confirming the suppositions in this study.In this analysis we will explore two hypotheses:

$HA1$—User organizations deliver a higher level of performance in ICSE than indifferent organizations;
$HB1$—Award winning organizations deliver a higher level of performance in ICSE than engaged organizations.

The first is that companies using the EMM achieved better satisfaction rates than indifferent ones during the period analyzed ($HA1$), a fact explored by Tutuncu and Kucukusta, see [23], which highlights customer satisfaction and image improvements, as one of the effects more significant among the companies that used the reference model and the second that awarded organizations had better satisfaction rates in relation to those engaged during the analyzed period ($HB1$).

At a significance level of 95% it showed a better performance of the companies that use the EMM versus the indifferent companies. One may conclude that there is a real benefit in adopting the EMM either from a social point of view or from a real or regulatory perspective. Also, the difference between the awarded and engaged companies was evaluated and confirmed the better results for the awarded companies.

In Figs. 3 and 4 are presented the ICSE results obtained for the winning companies and for the indifferent ones, where the line represents the average Brazilian results.

For each company, the annual results are represented in bars, the bar yellow when the annual company results are lower than the average results line, and the bar blue when the annual company results are above the average results line. Similar graphs were obtained for the engaged companies. In that case, the graphical representations show better results for the engaged companies other than the indifferent ones, but still lower results than the winner companies. In order not to overload the text those are not here presented.

6 Discussion and Conclusions

In this work, the performance differences of ICSE of the Brazilian electric energy distribution companies were evaluated, over a 10-year period (2007–2016), respecting the EMM adoption for different levels of companies compromise.

In the first step, the performance difference between the user companies and the indifferent companies was confirmed through the validation of hypothesis $HA1$. This is in accordance with the study of Puay et al. [12], for which the best satisfaction results were found for companies using excellence models.

Validation of the second hypothesis, $HB1$, confirms that award-winning organizations outperform participating organizations in ICSE outcomes. Thus, the results support studies such as that of Escrig and Menezes [24], who point to an increased correlation between managerial maturity and results for organizations using this model.

It is concluded, in this case, that the companies that made efforts to be rewarded had better results than the companies that were engaged. The effort of the engaged, however, still results in better performance in relation to the indifferent companies, allowing to reaffirm the beneficial character of the adoption of the EMM.

It is thus clear that the decision of the AEDC to promote the use of this model among its distributors in a massive way was the right decision. Its use makes the sector more robust and prepared to deal with the challenges imposed by the increase in consumer demand, restrictions from regulatory bodies and the dynamism of the world scenario. It is clear that the EMM has brought a new level of management to the sector, capable of directly and significantly impacting consumers.

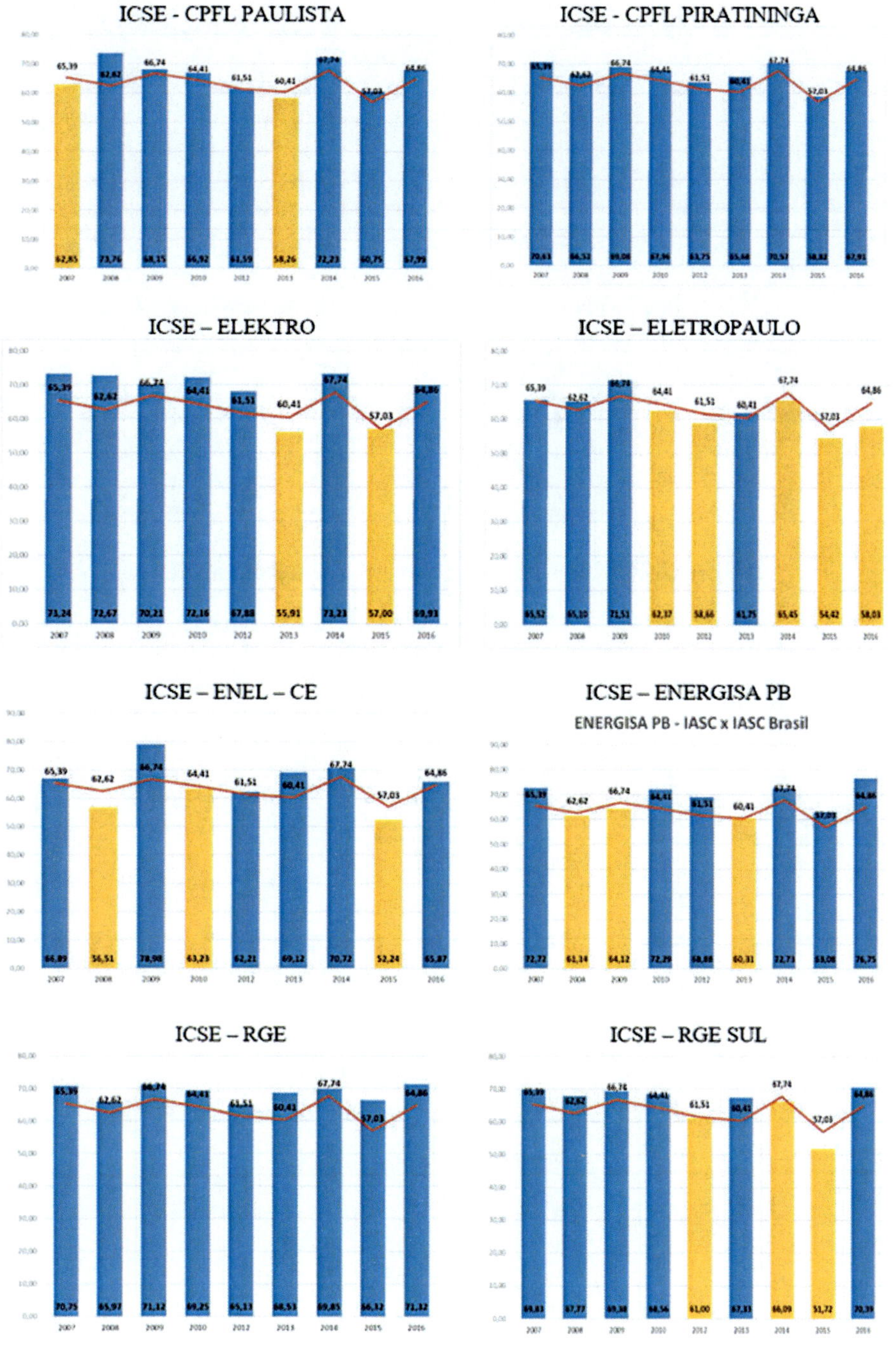

Fig. 3 Winner companies

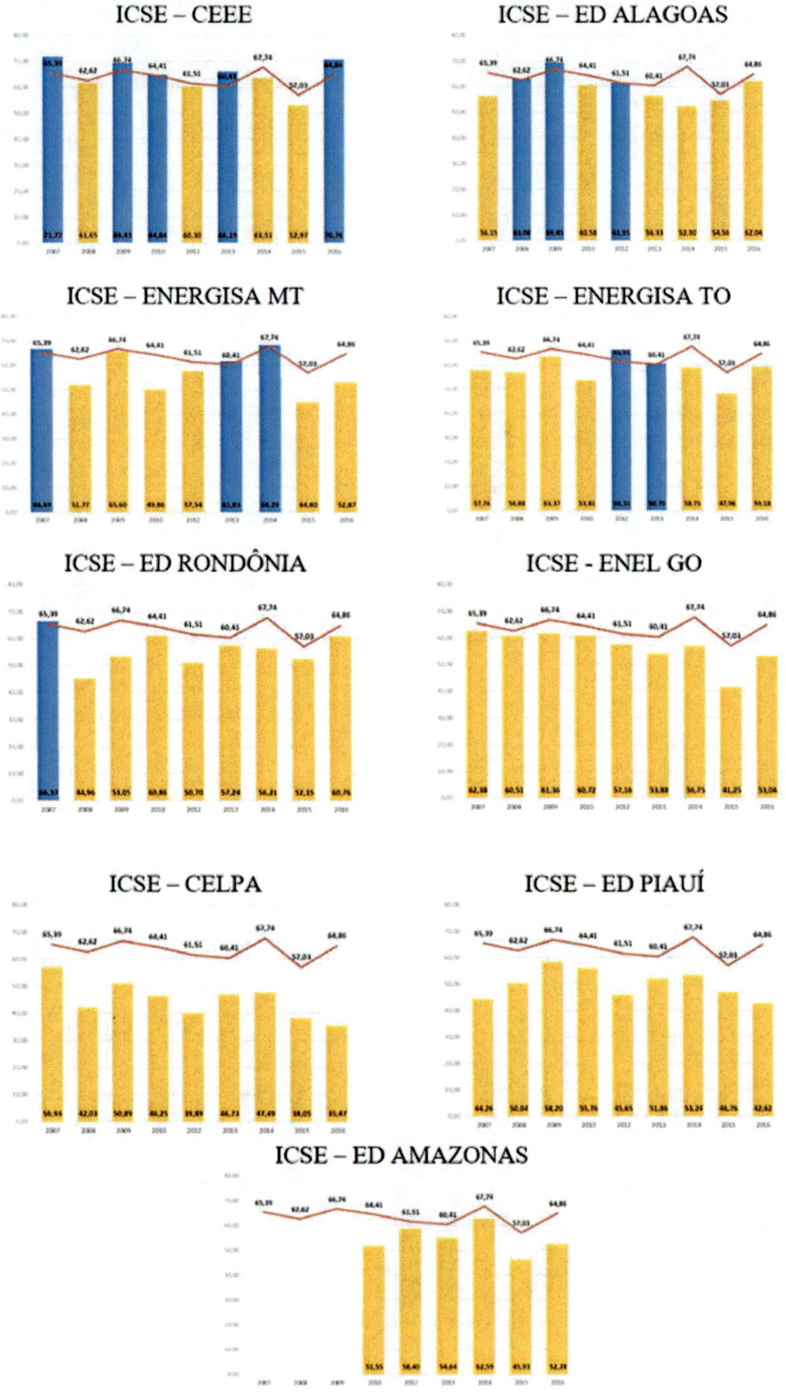

Fig. 4 Indifferent companies

The conclusions are consistent with the findings presented by Boulter et al. [2] and Corredor and Gofñi [3], and these credits come from companies that have achieved more positive results rather than those with poor results.

By adopting a combined and holistic approach, we also found that the higher the management maturity of a company, the better it can perform on the ICSE. Therefore, we can demonstrate the positive effects generated by using the reference model in the electrical energy sector on a large scale during the analysis period.

Appendix: List of Acronyms

AEDC Association of Electric Power Distribution Companies (Association representing Brazilian electricity distributors)

EMM Excellence in Brazilian Management Model (Management Maturity model used for the National Quality Prize)

GS Global Score (Score assigned by NQP to companies that submitted to the evaluation using the EMM)

ISCE Index of Consumer Satisfaction Evaluation (Survey promoted by the Brazilian regulatory agency annually)

NAEE National Agency of Electric Energy (Brazilian Regulatory Agency)

NQF National Quality Foundation (Brazilian institution that owns the EMM and promotes the NQP)

NQP National Quality Prize (Recognition offered to companies that have excellent management, similar to Deming Prize (Japan), the Malcolm Baldrige National Quality Award (USA) and the European Quality Award)

References

1. Baltazar, A.C.: Qualidade da energia no contexto da reestruturação do setor elétrico brasileiro. Dissertação de Mestrado. Escola Politécnica. Faculdade de Economia e Administração (2007)
2. Boulter, L., Bendell, T., Dahlgaard, J.J.: Total quality beyond North America: a comparative analysis of the performance of European excellence award winners. Int. J. Oper. Prod. Manag. **33**(2), 197 (2013)
3. Corredor, P., Gofii, S.: Quality awards and performance: is there a relationship? TQM J. **22**(5), 529–538 (2010)
4. Engel, J.F., Blackwell, R.D., Miniard, P.W.: Consumer Behavior. Dryden Press, Forth Worth (1993)
5. Buccelli, D.O., Costa Neto, P.L.O.: Prémio Nacional da Qualidade: Gestão da qualidade ou qualidade da gestão? Trabalho apresentado no XXXIII Encontro Nacional de Engenharia de Produção. A Gestão dos Processos de Produção e as Parcerias Globais para o Desenvolvimento Sustentável dos Sistemas Produtivos Bahia, Brasil 08 a 11 de outubro (2013)
6. Carrasco, A.: Dez anos de estudos sobre o impacto do uso modelos de excelência na qualidade do fornecimento (DEC e FEC) e satisfaqão de clientes no setor de distribuiqão de energia elétrica brasileiro. Dissertaqão de Mestrado. Gestäo. ISCTE – Instituto Universitário de Lisboa

(2018) CNI. Confederaqão Nacional da Indüstria: Retratos da sociedade brasileira – serviqos püblicos, tributaqão e gasto do governo. Indicadores CNI **5**(33), 1–14 (2016)
7. Marchetti, R. , Prado, P.H.M.: Um tour pelas medidas de satisfação do consumidor. Rev. Adm. Empresas **41**(4), 56–67 (2001)
8. Marchetti, R. , Prado, P.H.M.: Avaliaqão da satisfação do consumidor utilizando o método de equações estruturais: Um modelo aplicado ao setor elétrico brasileiro. Rac **8**(4), 9–32 (2004)
9. NAEE – Agência Nacional de Energia Elétrica. Regulamento Prêmio Iasc. Despacho N° 2.502, (2017)
10. Oliver, R.L.: Measurement and evaluation of satisfaction processes in retailing settings. J. Retail. **57**(3), 25–48 (1981)
11. Oliver, R.L.: Satisfaction: A Behavioral Approach. McGraw-Hill, Nova York (1997)
12. Puay, S.H., Tan, K.C., Xie, M., Goh, T.N.: A comparative study of nine national quality awards. TQM Mag. **10**(1), 30–39 (1998)
13. Miguel, P.A.C., Morini, C., Pires, S.R.I.: Um caso de aplicação do Prémio Nacional da Qualidade. TQM Mag. **16**(3), 186–193 (2004)
14. Cardoso, R., Cormack, A.M., Delesposte, J.E., Nascimento, M.K., Boechat, A.S.: O uso da ferramenta "metamodelo de gestão" na integração de múltiplos modelos de referência na modelagem da gestão organizacional. In: Trabalho apresentado no XIX Simpósio de Engenharia de Produção Sustentabilidade na Cadeia de Suprimentos, São Paulo, Brasil, 5 a 7 de Novembro (2012)
15. Calvo-Mora, A., Navarro-Garcia, A., Perianez-Cristobal, R.: Project to improve knowledge management and key business results through the EFQM. Int. J. Project Manag. **33**(8), 1638–1651 (2015)
16. Khoo, H.H., Tan, K.C.: Managing for quality in the USA and Japan: differences between the MBNQA, IDP, and JQA. TQM Mag. **15**(1), 14–24 (2003)
17. Tan, K.C.: A comparative study of 16 national quality awards. TQM Mag. **14**, 165–171 (2002)
18. NQF – Fundação Nacional da Qualidade Critérios de Excelência. Melhores em Gestão. Instruções para candidatura 2018. São Paulo (2018)
19. Doeleman, H.J., Ten Have, S., Ahaus, C.T.B.: Empirical evidence on applying the European foundation for quality management excellence model, a literature review. Total Qual. Manag. Bus. Excell. **25**(5–6), 439–460 (2014)
20. Vergara, S.C.: Projetos e Relatórios de Pesquisa em Administração. Atlas, São Paulo (1998)
21. Pino, A.F.: A questão da normalidade: uma revisão. Rev. Econ. Agrícola **61**(2), 17–33 (2014)
22. Martins, G.D.: Estatistica Geral e Aplicada. Atlas, Säo Paulo (2001)
23. Tutuncu, O., Kucukusta, D.: Relationship between organizational commitment and EFQM business excellence model: a study on Turkish quality awardwinners. Total Qual. Manag. **18**(10), 1083–1096 (2007)
24. Escrig, A.B., Menezes, L.M.: What characterizes leading companies within business excellence models? An analysis of "EFQM Recognized for Excellence" recipients in Spain. Int. J. Prod. Econ. **169**, 362–375 (2015)

A Statistical Boost to Assess Water Quality

Clara Cordeiro, Farhat-Un-Nisá Bajwa, and Sónia Cristina

Abstract Water quality in coastal and oceanic zones promotes various benefits for the regional economy, socio-cultural values, and biodiversity. Chlorophyll-*a* (Chl-*a*) is one of the most widely used water quality indicators. Monthly time series of Chl-*a* from 1998 until 2020 from two sites on the south coast of Portugal, Guadiana and Sagres, are used. Sagres is characterized by strong seasonality, and Guadiana with a weaker seasonal variation. A comparison between the months shows that Sagres is statistically significant when comparing the winter months with the early spring/summer months. Guadiana shows higher Chl-*a* values than Sagres but fewer changes between the months. A decrease in Chl-*a* concentration is detected in Guadiana, and its magnitude is obtained. Conversely, no monotonic trend is detected in Sagres. The approaches used must be viewed as exploratory. However, the findings might contribute to new ideas on the good environmental status of marine waters.

Keywords Chl-*a* satellite time series · Correlated Seasonal Kendall · Kruskal-Wallis · Trend magnitude

---

C. Cordeiro (✉)
Faculdade de Ciências e Tecnologia (FCT), Universidade do Algarve (UAlg), Faro, Portugal

CEAUL - Centro de Estatística e Aplicações, Faculdade de Ciências, Universidade de Lisboa, Lisboa, Portugal
e-mail: ccordei@ualg.pt

F.-U.-N. Bajwa
FCT, UAlg, Faro, Portugal
e-mail: a66695@ualg.pt

S. Cristina
CIMA-Centre for Marine and Environmental Research/ARNET-Aquatic Research Network, UAlg, Faro, Portugal
e-mail: sccristina@ualg.pt

© The Author(s), under exclusive license to Springer Nature Switzerland AG 2023
C. P. Kitsos et al. (eds.), *Statistical Modelling and Risk Analysis*,
Springer Proceedings in Mathematics & Statistics 430,
https://doi.org/10.1007/978-3-031-39864-3_5

1 Introduction

The high urbanised zones in the coastal regions and maritime economic activities are increasing globally, consequently causing pressures on coastal ecosystems [1] and also influencing the intermediate waters [2]. Anthropogenic pressures, both direct (e.g. fishing, deep-sea mining, aquaculture, ballast-spread invasive species) and indirect interactions from atmosphere-ocean (e.g. climate change and ocean acidification) and from land-ocean (e.g. effluents from industrial, urban and agriculture sources), affect global coastal ecosystems, regardless of their conservation status [3]. This can result in a loss of important ecosystem services, due to habitat reduction and associated biodiversity loss. Therefore, it is essential to monitor the water quality of marine waters to assess the effect of anthropogenic and natural inputs from nutrients on these waters, which can promote the occurrence of harmful algal blooms putting at risk human health and the coastal economic activities [4]. Furthermore, the increase in human activities, coupled with climate change, could exacerbate these effects. To protect these important pelagic habitats, it is crucial to comprehend the impact of pressures on marine areas [5]. In this way the European Union assess coastal ecosystems under the Marine Strategy Framework Directive 2008/56/EC (MSFD) and ensure that functions of pelagic habitats remain and achieve the Good Environment Status (GES; Commission Decision (EU) 2017/848) [3].

Currently, the water quality monitoring programs use in-situ data, which is limited by time and space and can be expensive. Alternatively, satellite ocean color remote sensing provides a cost-effective solution. Chlorophyll-*a* (Chl-*a*) concentration, which is a proxy for phytoplankton biomass, is one of the main water quality indicators that can be retrieved from space, allowing for monitor and evaluate the direct effect of nutrient enrichment in the marine waters. In this context, the MSFD uses the 90th percentile (P90) of Chl-*a* as an indicator for assessing GES and as a direct effect of eutrophication [4]. Coastal zones are among the most productive and diverse ecosystems, in Europe approximately 65% of the coastlines display signs of eutrophication [6]. Therefore, for a successful and efficient water quality monitoring programs in the marine waters of the Portuguese Exclusive Economic Zone (EEZ), the data retrieved from satellites are useful allies.

This work uses time series of Chl-*a* with a time horizon from January 1998 until December 2020 and from two sites off the coast of the Algarve, a region in the south of Portugal: Sagres and Guadiana. These time series have different stochastic behaviours, in which Sagres is characterised by a strong seasonal pattern, and, on the other hand, Guadiana presents a weak seasonality. Moreover, it was possible to investigate and identify the months that exhibit the most significant changes in the Chl-*a* concentration in the sites. In environmental monitoring, it may be useful to consider a method for detecting monotonic increasing or decreasing trends in water quality variables [7]. A modified seasonal Kendall test corrected for serial correlation among seasons (or months) was used, and the trend's magnitude was estimated. Although the methods presented here are viewed as exploratory, the results achieved might contribute to new ideas to be used in the evaluation of the status of marine water quality. The aim of this study is to analyze time series of Chl-*a* concentration using alternative statistical methods, providing additional perspectives on water quality assessment. This information is valuable to various stakeholders, including the Portuguese Environment Agency, coastal managers, and economic sectors like aquaculture, fisheries, and coastal tourism.

The paper is organised as follows. In Sect. 2, a brief description of EU Directives and the evaluation of waters' quality. In Sect. 3, the description of the statistical methodologies is followed by the case study in Sect. 4. Some concluding remarks will end this paper.

2 Some Background on EU Directives and Water Quality

Since 2000, the Member States of the European Union (EU) have committed to achieve a good ecological quality status of inland, coastal and transitional waters under the Water Framework Directive (WFD, 2000/60/EC). Due to this concern, European directives such as the WFD and MSFD [8] were developed to achieve and maintain these waters good ecological status and GES, respectively.

Coastal waters are the interface between land and the ocean. They are connected with waters from estuaries, rivers, harbours, and others. These coastal zones provide a range of economic, social, and ecological benefits. One of the most widely used water quality indicators is Chl-*a*, one of the elements to assess eutrophication (descriptor 5) of the MSFD. A commonly used descriptive statistics measure to assess the Chl-*a* concentration is the 90th percentile (P_{90}), because it is robust in the presence of outliers and extreme observations. This simple measure has been used in the WFD and MSFD [8]. The usual procedure is to determine the P_{90} and then see if the Chl-*a* values are below or above the reference/limit values established for the region under study. Reference conditions describes the reference mark against which current conditions are compared when assessing the status of water bodies [9]. Limit values are the benchmarks designed to alert when the current water conditions may be affected by external inputs (e.g. nutrients, pollutants, etc.).

In order to assess the GES of descriptor 5—eutrophication of the MSFD, the marine waters were delimited in three major areas in the Portuguese continental waters where the limits of the assessment areas were adopted from the WFD for coastal waters and were lengthen up to the EEZ boundaries [10]. Each of these areas are divided into smaller sub-area due to the geographic and oceanographic spatial heterogeneity of this wide region that will include coastal, intermediate and oceanic waters. In each sub-area, where applied the criteria and indicators to monitor and evaluate the state of the descriptor 5 (among them the Chl-*a* concentration as indicator) and where used the reference conditions and limit values of each indicator.

This work focuses on two sites and according to the areas delimited to evaluate the GES of the MSFD on the Portuguese continental coast these sites are inside the intermediate waters (waters between the outer limits of coastal waters and areas

Table 1 Thresholds for intermediate waters adapted from [12]

Site	Reference (R)	Limit (L)
Sagres	2.0 mg m^{-3}	3.0 mg m^{-3}
Guadiana	1.8 mg m^{-3}	2.7 mg m^{-3}

with a depth of less than 100 m deep) and each site is inside different sub-areas of the intermediate waters. Table 1 summarizes the reference conditions and the limit values that correspond to the sub-areas of the intermediate waters where the two sites are located. The limits for the reference condition and limit values for the intermediate waters were defined in the initial assessment of the MSFD for the Portuguese waters as described in [10, 11]. These values were reached based on the ecological characteristics, effect coastal upwelling and salinity regime of continental marine waters [12].

3 Statistical Utilities

3.1 Checking Independence

Autocorrelation measures and explains the internal correlation (dependence) between observations in a time series but shifts in time periods (lag). The Autocorrelation Function (ACF) is a widely used graphical approach for this purpose. If the ACF plot show autocorrelation at one or more lags, then the data is not independent. Another alternative is the Ljung-Box test [13], which checks if the autocorrelations are significantly different from a white noise series, i.e. a time series that shows no autocorrelation.

3.2 Comparing Seasons

Kruskal-Wallis test (KW) [14] is a rank-based nonparametric test used to investigate whether the medians of more than two independent groups (or seasons) of an independent variable with homogeneous variance are significantly different. The nonparametric Levene's test is used to test the homogeneity of variance [15]. In case of variance heterogeneity, the KW is affected, and the Welch (or Satterthwaite) approximation to the degrees of freedom is used [16]. The null hypothesis of the KW is that the group population medians are equal, and the alternative hypothesis is that at least two groups will differ. If the null hypothesis is rejected ($p\text{-}value < \alpha$), the result of the KW test is statistically significant, and a post hoc analysis is performed to determine which groups differ from each other group. A test that has been frequently used in such comparisons is Dunn's test [17]. The multiple pairwise comparisons are performed by the Dunn [18] with the p-values adjusted by a false discovery rate method, the Holm method [19, 20].

3.3 Detecting Correlated Seasons

When dealing with quantitative variables, is it useful to observe whether there are any statistical relationships between them. Correlation measures give a numerical measure of the relationship between variables. The most commonly used correlation measures are Pearson and Spearman. The latter is the nonparametric variant of the Pearson correlation. The Spearman's rank correlation coefficient has several advantages over Pearson, such as it does not depend on the normal distribution and is robust against outliers.

3.4 Detecting and Estimating Trends

Trend analysis in time series has been applied in several research fields, such as in water quality [21]. However, some time series characteristics complicate any analysis, such as skewness, non-normal data, presence of seasonality, serial correlation, outliers, and missing observations [21]. The nonparametric trend test Mann-Kendall (MK) [22, 23] was developed to deal with this type of data. Let $\{y_1, \cdots y_n\}$ be a sequence of observations indexed by time at regular time intervals. The hypotheses of the MK test are H_0 : no monotonic trend vs H_1 : either upward or downward trend. The test statistic S is defined as

$$S = \sum_{k=1}^{n-1} \sum_{j=k+1}^{n} sgn(y_j - y_k), \tag{1}$$

where $sgn(y_j - y_k)$ is calculated as described below

$$sgn(y_j - y_k) = \begin{cases} 1 & \text{if } y_j - y_k > 0 \\ 0 & \text{if } y_j - y_k = 0 \\ -1 & \text{if } y_j - y_k < 0 \end{cases}$$

and n is the sample size, y_k and y_j are observations at time instances k and j, with $k, j \leq n$ and $k \neq j$ [21]. Under the null hypothesis, the distribution of the test statistic S (1) is asymptotically normally distributed. See [21] for more details.

Hirsch et al. [21] developed a modified MK test to detect a trend in time series with seasonality designated by Seasonal Kendall (SK). The MK test statistic is calculated for each season i as follows

$$S_i = \sum_{k=1}^{m-1} \sum_{j=+1}^{m} sgn(y_{ij} - y_{ik}), \quad i = 1, \cdots, p \tag{2}$$

where m is the number of years, and p is the number of seasons (e.g. p=12 for monthly data). Then S_i (2) and its variance $Var(S_i)$ are calculated and the results combined to define the SK test statistics, S^*, as follows

$$S^* = \sum_{i=1}^{p} S_i \quad Var(S^*) = \sum_{i=1}^{p} Var(S_i) + \sum_{i=1}^{p} \sum_{h=1,h\neq i}^{p} Cov(S_i, S_h), \tag{3}$$

where S_i is given by (1) for the ith season (or month) and p is the total number of seasons. The SK test [21], assume that the statistics $S_1, \cdots, S_p$ are independent, so all the covariances terms in (3) are zero.

The SK test is robust in the case of seasonality and departures from normality but not against dependence between the seasons. To overcome this limitation, Hirsch and Slack [24] develop a test that performs better in case of correlation (Sect. 3.3) between the seasons, named here as Correlated Seasonal Kendall (CSK). This test uses estimates of the covariances between two seasons when calculating the variance in (3) and then corrects the normal approximation in the following expression:

$$Z^* = \begin{cases} \frac{S^*-1}{\sqrt{Var(S^*)}} & \text{if } S^* > 0 \\ 0 & \text{if } S^* = 0 \\ \frac{S^*+1}{\sqrt{Var(S^*)}} & \text{if } S^* < 0. \end{cases} \tag{4}$$

If the null hypothesis is rejected, an upward (downward) trend is detected if the value of S^* (4) is positive (negative). See [24] for more details.

The Slope Estimator A trend is a measure of the monotonic change during the time horizon. In contrast, the trend slopes represent the median rate of change for the selected period [7]. If a trend exists, its strength/magnitude (as a change per unit of time) and direction (negative/positive) can be estimated using the nonparametric Sen's method [25]. The magnitude is expressed as a slope, and the idea for estimating the slope is simple. For all pairs of time instances (i,j) where $i < j$, the slopes Q_i are calculated using the following expression

$$Q_i = \frac{y_j - y_i}{j - i}, \quad i = 1, \cdots, n-1, \quad j = 2, \cdots, n, \tag{5}$$

where y_j and y_i are observations at instances j and i respectively, and $i < j$. After computing all the Q_i slopes, the Sen's slope (Q) estimate is the median of all these values, i.e. $Q = median(Q_i)$, where $i = 1, \cdots, n-1$. In the case of a seasonal pattern in the data, the slopes in (5) are calculated between pairs of observations within the same season. The seasonal Sen slope estimator is then the median of all these individual slopes. The advantage of this method is that the estimate is robust in the presence of extreme observations and seasonal patterns [21].

3.5 Additional Notes and Software

The present research was conducted using a level of significance of $\alpha = 5\%$. The analysis was performed using R software version 4.2.1 [26] and packages including corrplot [27], forecast [28], ggplot2 [29], ggpubr [30], ggstatsplot [31], Rcmdr [32], trend [33], and wql [34].

4 Analysing Water Quality in Two Study Sites

4.1 Study Sites Description

In this case study, two sites in the south coast of Portugal (Algarve region), are described below (see Fig. 1):

Site 1: Off Sagres, located in the extreme west part of the region, lies in the transitional zone between the west and the south coasts of Portugal;

Site 2: Off Guadiana, located in the east part of the south Portuguese coast, is subject to the influence of Guadiana Estuary, a river between Portugal and Spain.

Monthly time series of Chl-*a* were obtained from multiple ocean colour sensors[1] onboard different satellite missions with 1 km of resolution were downloaded from the E.U. Copernicus Marine Environmental Monitoring Service[2] on the 26th February 2022 [35]. The time period covered is from January 1998 to December 2020, and for both study sites, located 3 km from the coast (see Fig. 1) was considered to minimize problems related to the contamination of the coast in the Chl-*a* concentration.

The time series plots (Fig. 2), show a significant difference between the two study sites: the Chl-*a* levels in the Guadiana are much higher than in Sagres, mainly due to its proximity to the Guadiana estuary. Another observation is that the Chl-*a* concentration in Guadiana has been decreasing over the 23-year period. Regarding the **R** and **L** thresholds described in Table 1, Sagres is classified as a "High state" since the Chl-*a* values are generally reported below the thresholds,

[1] Sea-viewing Wide Field-of-view Sensor (SeaWiFS), Moderate-Resolution Imaging Spectroradiometer (MODIS), Medium Resolution Imaging Spectrometer (MERIS), Visible Infrared Imaging Radiometer Suite (VIIRS) and Ocean and Land Colour Instrument (OLCI).

[2] Database: OCEANCOLOUR_ATL_CHL_L4_REP_OBSERVATIONS_009_091.

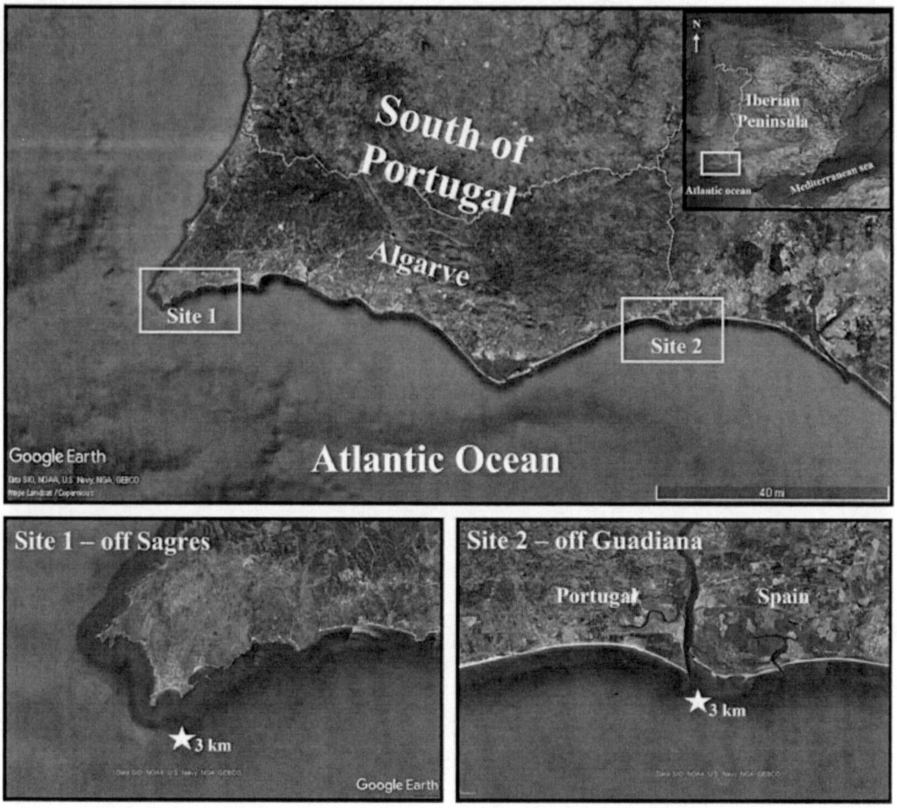

Fig. 1 Geographical location of the sites. Source: Satellite images from Google Earth

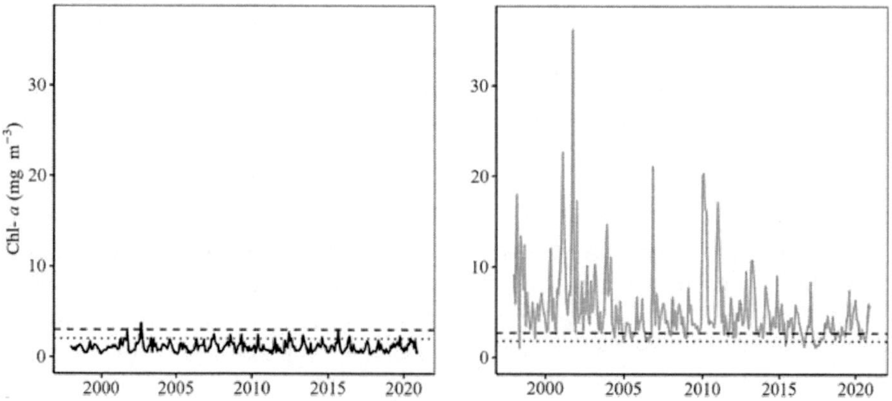

Fig. 2 Time series plots off Sagres (on the left) with the thresholds of **R** = 2.0 mg m^{-3} (dashed line) and **L** = 3.0 mg m^{-3} (dotted line), and off Guadiana (on the right) with **R** = 1.8 mg m^{-3} and **L** = 2.7 mg m^{-3}

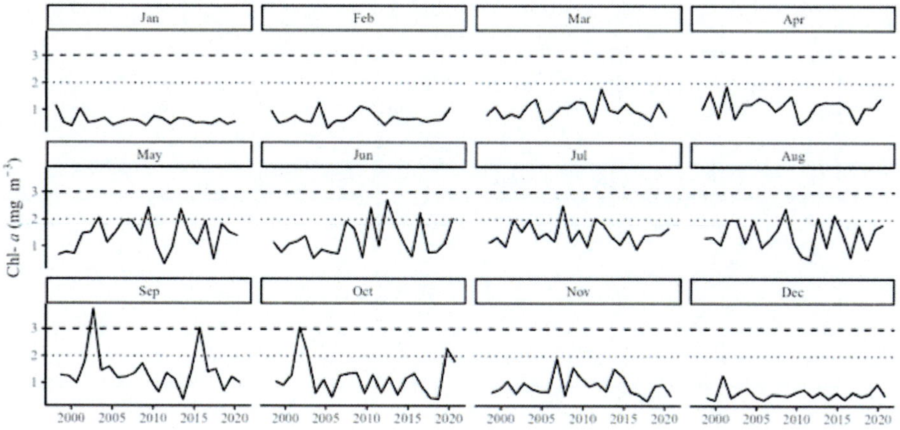

Fig. 3 Yearly variability off Sagres by month with the thresholds of **R** = 2.0 mg m^{-3} (dashed line) and **L** = 3.0 mg m^{-3} (dotted line)

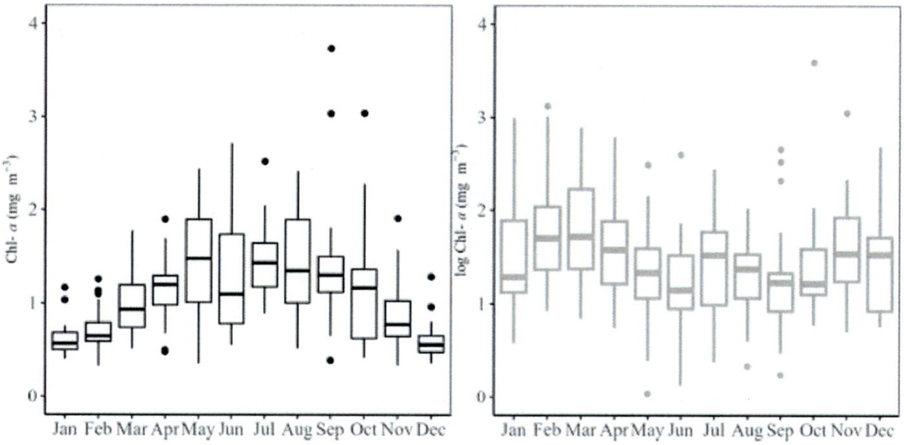

Fig. 4 Monthly variability off Sagres (on the left) and Guadiana (on the right)

as seen in Fig. 2. However, there are some exceptions, where some of the years recorded values were higher than the **R** threshold during the months of spring and the end of summer, and for the **L** threshold during October 2001 (3.04 mg m^{-3}), September 2002 (3.74 mg m^{-3}) and September 2015 (3.04 mg m^{-3}), as shown in Fig. 3. In the case of Guadiana, both thresholds (Table 1) are frequently exceeded (Fig. 2).

In time series analysis, it is common to look at patterns in the data, such as seasonality. Through the autocorrelation plot (Sect. 3.1), it is possible to observe a strong seasonal pattern in Sagres (Fig. 5), with higher Chl-*a* values in spring and summer, as seen in Fig. 4, which is consistent by the fact that the phytoplankton growing season is considered to be from February to October [36]. Concerning

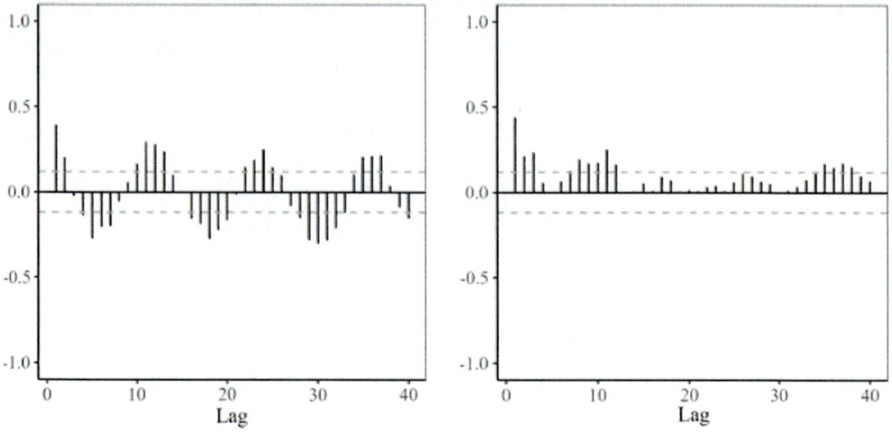

Fig. 5 ACF for Sagres (on the left) and Guadiana (on the right)

Guadiana, it is possible to detect small positive correlation coefficients in the ACF plot (Fig. 5) at multiples of 12, which indicates a weak seasonal pattern. Although with a weak seasonality, the Guadiana site presents higher values during the winter (Nov/Dec) and early spring (Feb/Mar), as seen in Fig. 4. This result might be related to the geographic location as it is close to the river estuarine and it has a dam upstream [37]. Therefore, it is reasonable to associate these high levels of Chl-*a* with the rainy season and dam discharges.

4.2 *Statistical Changes Between the Months*

Parametric methods rely on underlying assumptions about the probability distribution, while nonparametric methods do not require such assumptions. These methods are robust enough to analyze water quality data, which often have missing values, outliers and non-normal data. The nonparametric Kruskal-Wallis test was used to investigate whether there was a significant difference between the seasons (months). Although nonparametric methods are distribution-free, fewer assumptions about the data need to be checked, such as independence. The hypothesis test Ljung-Box (LB) (Sect. 3.1) is used to measure the observations' dependence, whether autocorrelation is present or not in the time series. The LB results show no autocorrelation since the null hypothesis is not rejected ($p\text{-}value > \alpha$) for each month and each site. Then the Levene test was applied on each site revealing that Guadiana assumes equal variances between the months ($p\text{-}value^{G}_{Levene} = 0.577$), that is, the null hypothesis is not rejected at 5% level of significance. In the case of Sagres, the homogeneity of variances fails, and so an opposite conclusion was reached, heterogeneity of variances ($p\text{-}value^{S}_{Levene} = 1.535\ e^{-05}$). Thus, the KW was performed considering the Welch (or Satterthwaite) approximation, as explained in Sect. 3.2. The KW was

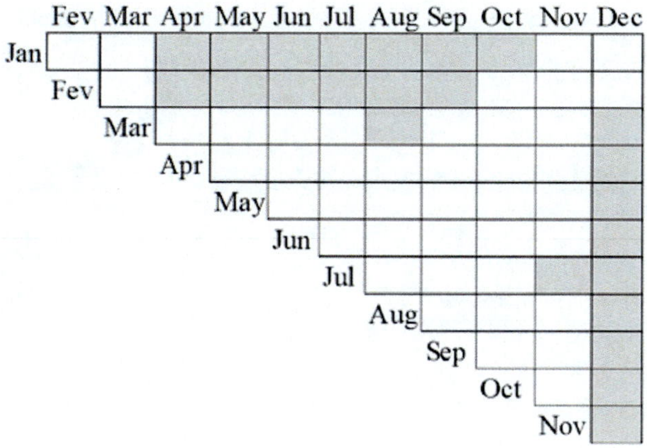

Fig. 6 Nonparametric multiple comparison results at Sagres (p-value $<$ 5%)

conducted to examine the monthly changes in Chl-a concentrations at 3 km from the coast. For both sites, the null hypothesis is rejected ($p\text{-}value^{G}_{KW} = 0.0034$, $p\text{-}value^{S}_{KW} < 2.2\ e^{-16}$), showing evidence of a significant difference between the months. Furthermore, the Dunn test was applied to examine which months present the most significant changes, and the Holm method was used in the p-adjustment. Sagres has shown several statistically significant changes between the winter months (Nov, Dec, Jan, Feb) and the others, as seen in Fig. 6. In the spring/summer months, there is an increase in the phytoplankton on this site, as mentioned by several studies [36]. Conversely, Guadiana has only detect changes between the pairs (Feb, Mar) ($p\text{-}value_{Holm\text{-}adj}$=0.04).

4.3 *Investigating Temporal Trends*

Another interesting issue in water quality monitoring is investigating whether Chl-a concentration has increased/decreased along the considered time horizon. As explained in Sect. 3.3, the time series have characteristics to which the parametric approaches fail. So, a nonparametric trend test, with seasonal modification, from the Mann-Kendall family is used. However, in the presence of correlation, the correlated version of the SK is used, as described in Sect. 3.3. Therefore, the nonparametric Spearman correlation coefficient and its significance were achieved, as seen in Fig. 7. In the case of Guadiana, only a few relationships were observed based on the moderate correlation coefficients. The opposite situation is observed at the Sagres site with several positive and high correlation coefficients between the months,

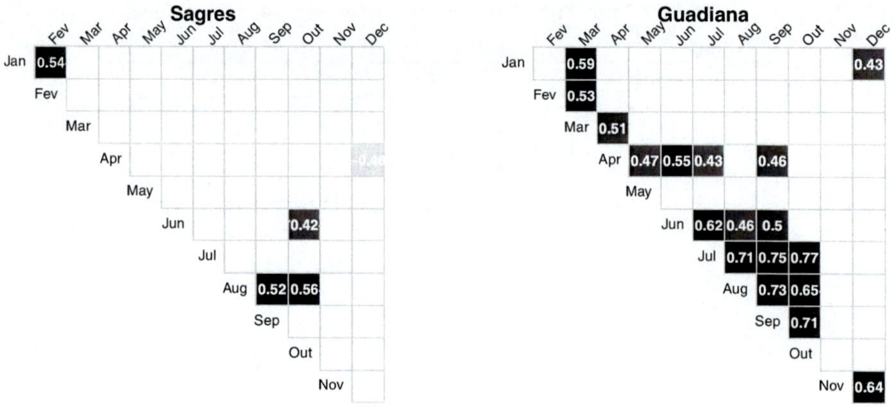

Fig. 7 Spearman correlation coefficients (only p-values statistically significant)

Table 2 Correlated Seasonal Kendall test results

Site	Z^*	p-Value	Decision
Sagres	−0.1580	0.8745	No trend
Guadiana	−3.1285	**0.0018**	Trend (–)

namely Jul/Aug/Sep. According to the presence of correlation between the seasons (months), the CSK test was performed, and the results are shown in Table 2. As observed in this Table, no statistically significant trend was detected in Sagres. However, in the case of Guadiana, a significantly decreasing trend was detected. For this site, the magnitude of the trend was calculated using the seasonal Sen Slope estimator, as described in Sect. 3.3. Over the 23 years, the Chl-*a* concentration have decreased significantly at an approximate rate of $-1.36\,\text{mg m}^{-3}$/year at 3 km. Moreover, Fig. 8 show the results of the MK test and Sen's slope estimate for each month. In general, it is possible to observe a decreasing Chl-*a* trend, particularly in the spring/autumn months. These findings are related to the drought affecting the Algarve for some years and, consequently, the lower occurrence of river discharges during these months.

5 Concluding Remarks and Thoughts

This work provides an overview of some statistical approaches to analysing one of the most commonly used water quality indicators, Chl-*a*. Declining coastal water quality has implications for various sectors, such as health, offshore aquaculture, tourism and other sectors that rely on ecosystem services. Effective monitoring programs are therefore essential to maintain and achieve a GES in which marine and coastal waters benefit from satellite remote sensing data due to their spatial and temporal coverage.

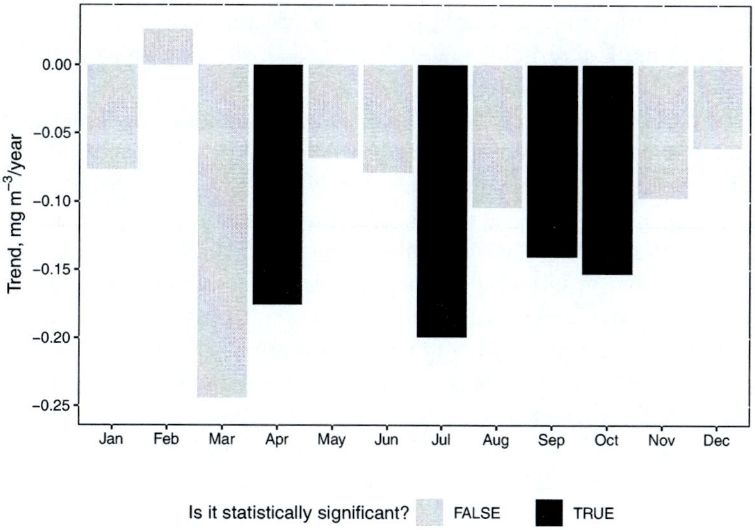

Fig. 8 MK test and Sen's slope estimates at Guadiana

In this study, 23 years of monthly satellite time series of Chl-*a* are considered to monitor the water quality at two sites, Sagres and Guadiana, 3 km off the southern coast of Portugal. **Sagres** is characterized by a strong seasonal pattern with high Chl-*a* concentrations during spring and end of summer, yet it is still below $\mathbf{R} = 2.0\,\mathrm{mg\,m^{-3}}$ according to the MSFD. Moreover, it is possible to define two distinct periods: October until March, with lower Chl-*a* concentrations, and April until September with higher levels, which are consistent with the upwelling season events [36]. Therefore, based on the strong seasonality and the changes between the months, this might suggest that the **R** and **L** thresholds could be adjusted to this monthly variation. A weak seasonal pattern characterises **Guadiana**, in which the Chl-*a* concentration values often exceeded the **R** and **L** thresholds proposed for this parameter for the implementation of the MSFD. However, in recent years the Chl-*a* have been decreasing and in some cases the values were recorded below both thresholds. One of the reasons for this decrease might be associated with the Algarve region suffering from periods of low precipitation and to the reduction of dam discharges into the Guadiana river, consequently important events, as the ocean fertilisation, would happen in a smaller magnitude in these intermediate waters. For this reason, identifying sites influenced by river discharges, such as off Guadiana estuary, empower the management and conservation of these important ecosystem in the future. Another approach was detecting and estimating the magnitude of temporal trends in water quality. The Seasonal Kendall test was applied to detect temporal patterns since it is insensitive to the presence of seasonality. Trend analysis reveal a decreasing trend, especially in the spring/summer months. When comparing by season (month), the rate of change is bigger in July. Overall, this research study highlights that Sagres maintain Chl-*a* concentrations under the

MSFD thresholds, and Guadiana, besides the decrease of Chl-*a* values in recent years, the **R** and **L** thresholds are often exceeded, showing some risk regarding to the Chl-*a* concentrations within this site.

To the best of the authors' knowledge, no other studies using Kruskal-Wallis and Seasonal Kendall to analyse the coastal water quality were found for these study sites. Moreover, this study contributes to the literature by describing the temporal evolution of the Chl-*a* as one of the indicators of water quality in these two study sites at 3 km. Furthermore, it is essential to develop methodologies to maintain water quality levels and ensure that the population can continue to benefit from ocean resources and ecosystem services. In future work, it is planned to extend this study to more stations, covering all continental Portuguese coast. In addition, it is also planned to include the data related to the river discharges and investigate the effect of discharges on the Chl-*a* concentration, i.e. its association with the decreasing trend and other parameters, such as dissolved oxygen and nutrients used as indicators to assess the environmental descriptor 5 (eutrophication) of the MSFD.

Acknowledgments The authors would like to thank two referees for their comments and suggestions, which improved certain aspects and the presentation of this paper. Clara Cordeiro is partially financed by national funds through FCT—Fundaçõo para a Ciência e a Tecnologia under the project UIDB/00006/2020. Farhat-Un-Nisa Bajwa would like to thank the CEAUL for support through Grant UIDP/00006/2020 funded by FCT through national funds. Sónia Cristina is financed through the FCT under the grant: CEECIND/01635/2017 and would like to acknowledge the financial support of FCT to CIMA through project UIDP/00350/2020 and through project LA/P/0069/2020 granted to the Associate laboratory ARNET. The authors also thank M. Rosário Ramos for discussions about the nonparametric trend tests.

References

1. Cristina, S., Icely, J., Costa Goela, P., Angel DelValls, T., Newton, A.: Using remote sensing as a support to the implementation of the European Marine Strategy Framework Directive in SW Portugal. Cont. Shelf Res. **108**, 169–177 (2015)
2. MAMAOT: Marine Strategy for the subdivision of Continent. Update on the 2nd Cycle Report (D). Marine Strategy Framework. The General Inspection of Agriculture, Sea, Environment and Spatial Planning, Portugal, pp. 359 (2020)
3. Magliozzi, C., Palma, M., Druon, J.-N., Palialexis, A., Abigail, M.Q.G., Ioanna, V., Rafael, G.-Q., Elena, G., Birgit, H., Laura, B., Felipe, A.L.: Status of pelagic habitats within the EU-marine strategy framework directive: proposals for improving consistency and representativeness of the assessment. Mar. Policy **148**, 105467 (2023). https://doi.org/10.1016/j.marpol.2022.105467
4. Brito, A.C., Garrido-Amador, P., Gameiro, C., Nogueira, M., Moita, M.T., Cabrita, M.T.: Integrating in situ and ocean color data to evaluate ecological quality under the water framework directive. Water **12**(12), 3443 (2020). https://doi.org/10.3390/w12123443
5. Rodrigues, M., Rosa, A., Cravo, A., Jacob, J., Fortunato, A.B.: Effects of climate change and anthropogenic pressures in the water quality of a coastal lagoon (Ria Formosa, Portugal). Sci. Total Environ. **780**, 146311 (2021)
6. Russ, J., Zaveri, E., Desbureaux, S., Damania, R., Rodella, A.S.: The impact of water quality of GDP growth: evidence from around the world. Water Security **17**, 100130 (2022)

7. Mozejko, J.: Detecting and estimating trends of water quality parameters. In: Water Quality Monitoring and Assessment, InTech, pp. 95Ű120 (2012). http://dx.doi.org/10.5772/33052
8. Directive 2008/56/EC of the European Parliament and of the Council of 17 June 2008. Official J. **164**, 19–40 (2008)
9. Poikane, S., Alves, M., Argillier, C., van den Berg, M., Buzzi, F., Hoehn, E., Hoyos, C., Karottki, I., Laplace-Treyture, C., Solheim, L.A., Ortiz-Casas, J., Ott, I., Phillips, G., Pilke, A., Padua, J., Remec-Rekar, S., Riedmuller, U., Schaumburg, J., Serrano, L.M., Soszka, H., Tierney, D., Urbanic, G., Wolfram, G.: Defining chlorophyll a reference conditions in European lakes. Environ. Manage. **45**, 1286–1298 (2010)
10. Cabrita, M.T., Silva, A., Oliveira, P.B., Angélico, M.M., Nogueira, M.: Assessing eutrophication in the Portuguese continental exclusive economic zone within the European marine strategy framework directive. Ecol. Indic. **58**, 286–299 (2015)
11. MAMAOT: Marine Strategy for the subdivision of the Continent. Marine Strategy Framework. The General Inspection of Agriculture, Sea, Environment and Spatial Planning, Portugal, pp. 906 (2012)
12. IPMA: 2nd cycle report of good environmental status assessment of marine waters in the mainland and extended to Continental Shelf subdivisions. Instituto Português do Mar e da Atmosfera. Marine Strategy Framework Directive. Descriptor 5 - anthropogenic eutrophication, pp. 20 (2018)
13. Hyndman, R.J., Athanasopoulos, G.: Forecasting: Principles and Practice, 2nd edn. OTexts, Melbourne (2018). Accessed on 4th July 2022. https://otexts.com/fpp2/
14. Kruskal, W.H., Wallis, W.A.: Use of ranks in one-criterion variance analysis. J. Am. Stat. Assoc. **47**(260), 583–621 (1952)
15. Zar, J.: Biostatistical Analysis, 5th edn. Pearson Prentice Hall, Upper Saddle River (2010)
16. Dag, O., Dolgun, A., Konar, N.M.: Onewaytests: An R Package for One-Way Tests in Independent Groups Designs. R J. **10**(1) (2018). https://doi.org/10.32614/RJ-2018-022
17. Dinno, A.: Nonparametric pairwise multiple comparisons in independent groups using Dunn's test. Stata J. **15**(1), 292–300 (2015)
18. Dunn, O.J.: Multiple comparisons using rank sums. Technometrics **6**(3), 241–252 (1964)
19. Holm, S.: A simple sequentially rejective multiple test procedure. Scand. J. Stat. **6**, 65–70 (1979)
20. Wright, S.P.: Adjusted p-values for simultaneous inference. Biometrics **48**, 1005–1013 (1992)
21. Hirsch, R. Slack, J., Smith, R.: Techniques of trend analysis for monthly water quality data. Water Resour. Res. **18**, 107–121 (1982)
22. Mann, H.B.: Nonparametric tests against trend. Econometrica **13**, 245–259 (1945)
23. Kendall, M.G.: Rank Correlation Methods, 4th edn. Charles Griffin, London (1975)
24. Hirsch, R.M., Slack, J.R.: A nonparametric trend test for seasonal data with serial dependence. Water Resour. Res. **20**(6), 727–732 (1984)
25. Sen, P.K.: Estimates of the regression coefficient based on Kendall's tau. J. Am. Stat. Assoc. **63**(324), 1379–1389 (1968)
26. R Core Team. R: A language and environment for statistical computing. R Found for Statistical Computing, Vienna (2022). https://www.R-project.org/
27. Wei, T., Simko, V.: R package 'corrplot': visualization of a Correlation Matrix. R package version 0.92 (2021). Available from https://github.com/taiyun/corrplot
28. Hyndman, R., Athanasopoulos, G., Bergmeir, C., Caceres, G., Chhay, L., O'Hara-Wild, M., Petropoulos, F., Razbash, S., Wang, E., Yasmeen, F.: Forecast: forecasting functions for time series and linear models. R package version 8.17.0 (2022). https://pkg.robjhyndman.com/forecast/
29. Wickham, H.: ggplot2: Elegant Graphics for Data Analysis. Springer, New York (2016)
30. Kassambara, A.: ggpubr: 'ggplot2' Based Publication Ready Plots. R package version 0.4.0 (2022). https://CRAN.R-project.org/package=ggpubr
31. Patil, I.: Visualizations with statistical details: the 'ggstatsplot' approach. J. Open Source Softw. **6**(61), 3167 (2021). https://doi.org/10.21105/joss.03167
32. Fox, J., Bouchet-Valat, M.: Rcmdr: R Commander. R package version 2.8-0 (2022)

33. Pohlert, T: trend: Non-Parametric Trend Tests and Change-Point Detection. R package version 1.1.4 (2020). https://CRAN.R-project.org/package=trend
34. Jassby, A.D., Cloer, J.E.: wq: Exploring water quality monitoring data. R package version 1.0.0 (2022). https://CRAN.R-project.org/package=wq
35. E.U. Copernicus Marine Service Information. Accessed on 26th February 2022. https://doi.org/10.48670/moi-00074
36. Goela, P.C., Cordeiro, C., Danchenko, S., Icely, J., Cristina, S., Newton, A.: Time series analysis of data for sea surface temperature and upwelling components from the southwest coast of Portugal. J. Mar. Syst. **163**, 12–22 (2016)
37. Brito, A.C., Brotas, V., Caetano, M., Coutinho, T.P., Bordalo, A.A., Icely, J., Neto, M.J., Serôdio, J., Moita, T.: Defining phytoplankton class boundaries in Portuguese transitional waters: an evaluation of the ecological quality status according to the Water Framework Directive. Ecol. Indic. **19**, 5–14 (2012)

Time Series Procedures to Improve Extreme Quantile Estimation

Clara Cordeiro, Dora P. Gomes, and M. Manuela Neves

Abstract Although extreme events can occur rarely, they may have significant social and economic impacts. To assess the risk of extreme events, it is important to study the extreme quantiles of the distribution. The accurate semi-parametric estimation of high quantiles depends strongly on the estimation of some crucial parameters that appear in extreme value theory. Procedures that combine extreme value theory and time series modelling have revealed themselves as a nice compromise to capture extreme events. Here we study the estimation of extreme quantiles after adequate time series modelling. Using the R software, our approach will be applied to the daily mean flow discharge rate values of two rivers in Portugal.

Keywords Extreme value theory · High quantiles · Semiparametric estimation · Time series

---

C. Cordeiro (✉)
Faculdade de Ciências e Tecnologia, Universidade do Algarve, Faro, Portugal

CEAUL – Centro de Estatística e Aplicações, Faculdade de Ciências, Universidade de Lisboa, Lisboa, Portugal
e-mail: ccordei@ualg.pt

D. P. Gomes
Center for Mathematics and Applications (NOVA Math) and Department of Mathematics, NOVA FCT, Caparica, Portugal
e-mail: dsrp@fct.unl.pt

M. M. Neves
Instituto Superior de Agronomia, Lisboa, Portugal

CEAUL – Centro de Estatística e Aplicações, Faculdade de Ciências, Universidade de Lisboa, Lisboa, Portugal
e-mail: manela@isa.ulisboa.pt

© The Author(s), under exclusive license to Springer Nature Switzerland AG 2023
C. P. Kitsos et al. (eds.), *Statistical Modelling and Risk Analysis*,
Springer Proceedings in Mathematics & Statistics 430,
https://doi.org/10.1007/978-3-031-39864-3_6

1 Motivation and Introduction

Extreme values research plays an important role in several practical domains of applications, such as hydrology, environment, rainfall data, earthquake analysis, finance, and insurance, among others. The extreme value theory (EVT) deals with appropriate statistical models to estimate extreme quantiles and probabilities of rare events. Extreme value models were initially obtained through arguments that assumed an underlying process consisting of a sequence of independent and identically random variables. However, in many real situations, temporal independence is unrealistic, and a stationary setup is the most natural generalisation of a sequence of independent random variables. In the last decades, many signs of progress have been made in parameter estimation of extreme values in time series relevant to asymptotic results. However, for finite samples, limiting results provide approximations that can be poor.

One of the main obstacles in modelling extreme events in time series is that the distribution of extreme values is frequently non-normal and skewed. To address this, EVT offers a framework for modelling such distributions, using models like the generalised extreme value (GEV) distribution and the generalised Pareto distribution (GPD). Whenever we are interested in the extremes of a time series, a time series model for the complete process can be fitted to the data and then determine its extremal behaviour. Another approach consists in choosing a model for the process at extreme levels only and fit it to the extremes in the data. This alternative is attractive because models for extremes can be derived under very weak conditions on the process. Methods that combine extreme value theory and time series models are the most flexible ones and can capture extreme events. Some work can be mentioned [1–3] among others. In [3] EVT and time series models are combined to obtain more reliable extreme value parameter estimates. In classical time series modelling, a key issue is to determine statistically how many parameters have to be included in the model. However, special care must be given to extreme events in the series that need specific statistical procedures based on the behaviour of extremes. Some recent references discuss the application of EVT and ARIMA or exponential smoothing models in applied sciences, such as meteorological data modelling [4].

In this paper, the best time series model is chosen to fit the extremes in a data set, using some well-known accuracy measures. An adequate EVT model is fitted to the residuals obtained, and the time series is reconstructed, obtaining a 'replica' of the original one. In EVT analysis, a few estimators illustrate the application of the procedure that is developed.

There are many applied sciences, such as economics, finance, environment, medical sciences, climatology, etc., where these procedures must be used to estimate, among other measures, very high quantiles because of their impact on human life.

Our approach is applied to daily mean flow discharge rate values of two main rivers in Portugal, using the R software [5]. The paper proceeds as follows. Section 2 contains the main results that are the basis of the theoretical background on extreme value analysis, including the description and properties of the parameter estimators considered. Section 3 briefly overviews time series models and their linkage with Extreme Value Theory. The application to the real data is performed in Sect. 4. Some final remarks and future work is presented in Sect. 5.

2 Main Results in Extreme Values Analysis

Nowadays it is crucial to quantify risk assessment of the effects of climate change. Society, ecosystems, etc. tend to adapt to routine, near-normal conditions: these conditions tend to produce fairly minimal impacts. In contrast, unusual and extreme conditions tend to have much more substantial net impacts despite, by definition, occurring a much smaller proportion of the time. Statistical analysis of extreme values was traditionally applied to hydrology and insurance. There is a quite large variety of fields of application of extreme value theory, such as climatology, oceanography, environment, and biology. Unlike most traditional central statistical theory, which typically examines the usual (or the average) behaviour of a process, extreme value theory deals with models for describing unusual behaviour or rare events. The heart of extreme value theory is the reliable extrapolation of values beyond the observed range of sample data. Modelling rare events of univariate time series is an area of important research. In classical time series modelling, a key issue is determining statistically how many parameters must be included in the model. However, special care must be given to extreme events in the series that need specific statistical procedures based on the behaviour of extremes. Extreme value models were initially obtained through arguments that assumed an underlying process consisting of a sequence of independent and identically random variables. However, temporal independence is unrealistic in many situations where extreme value models are of great interest to be applied. A stationary setup is the most natural generalisation of a sequence of independent random variables. In the last decades, much progress has been made in parameter estimation of extreme values in time series, with relevance to asymptotic results.

EVT is the branch of probability and statistics dedicated to characterizing the very low or quite high values of a variable, the tail of the distribution. EVT had its beginnings in the early to the middle part of the twentieth century. Emil Gumbel was the pioneer in the applications of statistics of extremes. In Statistics of Extremes, [6], he presents several applications of EVT on real-world problems in engineering and meteorological phenomena.

Let us assume that we have a sample $(X_1, \ldots, X_n)$ of independent and identically distributed (iid) or possibly stationary, weakly dependent random variables from an unknown cumulative distribution function (cdf) F. The interest is focused in the distribution of the maxima, $M_n := \max(X_1, \ldots, X_n)$, for which we have

$$P(M_n \leq x) = P(X_1 \leq x) \ldots P(X_n \leq x) = F^n(x). \quad (1)$$

As n goes to ∞, the distribution F^n in (1) has a trivial limit: 0, if $F(x) < 1$ and 1, if $F(x) = 1$. So the idea for M_n was the same of *central limit theorem*: first subtract a n-dependent constant, then rescale by a n-dependent factor. The question is then whether one can find two sequences, $\{a_n\} \in R^+$ and $\{b_n\} \in R$ and a non-trivial limiting distribution function, G, such that

$$\lim_{n\to\infty} P\left((M_n - b_n)/a_n \leq x\right) = G(x).$$

First results on the G distribution are due to [7–9] and [10]. But were [11] and [12] who gave conditions for the existence of those sequences $\{a_n\} \in R^+$ and $\{b_n\} \in R$ such that, when $n \to \infty$ and $\forall x \in R$,

$$\lim_{n\to\infty} P\left(\frac{M_n - b_n}{a_n} \leq x\right) = \lim_{n\to\infty} F^n(a_n x + b_n) = \mathrm{EV}_\xi(x), \tag{2}$$

where EV_ξ is a nondegenerate distribution function, denoted as the Extreme Value cdf, and given by

$$\mathrm{EV}_\xi(x) = \begin{cases} \exp[-(1+\xi x)^{-1/\xi}], & 1+\xi x > 0 \text{ if } \xi \neq 0 \\ \exp[-\exp(-x)], & x \in R \quad\quad \text{if } \xi = 0. \end{cases} \tag{3}$$

When the above limit holds we say that F is in the domain of attraction (for maxima) of EV_ξ and write $F \in \mathcal{D}_{\mathcal{M}}(\mathrm{EV}_\xi)$.

2.1 Parameters of Interest

The shape parameter ξ, that appears in (3), is called extreme value index (EVI) and it is the primary parameter of interest in the whole extreme value analysis. However, (3) can also incorporate location (λ) and scale ($\delta > 0$) parameters, and in this case, the EV_ξ cdf is given by,

$$\mathrm{EV}_\xi(x; \lambda, \delta) \equiv \mathrm{EV}_\xi((x-\lambda)/\delta). \tag{4}$$

The parameter ξ it also the basis of other important parameters of extreme events, such as:

- a *high quantile* of probability $1-p$ (p small)

$$\chi_{1-p} := \inf\{x : F(x) \geq 1-p\},$$

$$\chi_{1-p} := \lambda - \frac{\delta}{\xi}\left[1 - \{-\log(1-p)\}^{-\xi}\right], \quad \xi \neq 0$$

- the *probability of exceedance* of a high level;
- the *return period* of a high level,
- the *right endpoint* of an underlying model F,

$$w_F := \{x \in R : F(x) < 1\}$$

2.2 Statistical Approaches in EVT

EVT has been developed under two frameworks. The first one is the parametric framework that considers a class of models associated with the limiting behaviour of the maxima, given in (2). The main assumption behind the parametric approach is that estimators are calculated considering the data following approximately an exact EV probability distribution function, defined by a number of parameters.

In the semi-parametric framework, the only assumption made is that the limit in (2) holds, i.e., that the underlying distribution verifies the extreme value condition. In this framework we do not need to fit a specific parametric model based on scale, shape and location parameters. Estimates are now usually based on the largest k order statistics in the sample, assuming only that the model F underlying the data is in $\mathcal{D}_{\mathcal{M}}(\text{EV}_\xi)$.

The parameter ξ is estimated, on the basis of k top observations, with k intermediate, i.e. such that $k = k_n \to \infty$ and $k/n \to 0$, as $n \to \infty$. However most estimators show a strong dependence on that value k. They usually present: a small bias and a high variance for small values of k; bias increases and variance decreases when k increases.

Some Semi-parametric Estimators of ξ Currently there are several different EVI-estimators. For illustrating our study we are going to consider two of them.

Let $X_{1:n} \leq X_{2:n} \leq \ldots \leq X_{n:n}$ denote the associated non-decreasing order statistics from the sample of size n. The most popular semi-parametric EVI-estimator is the Hill estimator, H, [13]. This estimator can be defined as the average of the log-excesses, $V_{ik} := \ln X_{n-i+1:n} - \ln X_{n-k:n}$, $1 \leq i \leq k < n$, above the high threshold $X_{n-k:n} > 0$,

$$\widehat{\xi}_n^H(k) \equiv H(k) := \frac{1}{k}\sum_{i=1}^{k} V_{ik}, \quad 1 \leq k < n. \tag{5}$$

There are several alternatives to the Hill estimator that are less sensitive to the choice of the level k. We shall consider the simplest class of reduced-bias EVI-estimators, the Corrected Hill (CH) estimator [14], defined by

$$\hat{\xi}_n^{CH}(k) \equiv \hat{\xi}_{\hat{\beta},\hat{\rho}}^{CH}(k) \equiv \mathrm{CH}(k) := \mathrm{H}(k)\left(1 - \frac{\hat{\beta}(n/k)^{\hat{\rho}}}{1-\hat{\rho}}\right), \quad 1 \leq k < n, \tag{6}$$

with $(\hat{\beta}, \hat{\rho})$ an adequate estimator of the vector of second-order parameters (β, ρ), (see [15] and [16] for estimators for ρ and β, respectively).

A Semi-parametric Estimator of Extreme Quantiles Suppose we have a sample $(X_1, X_2, \ldots, X_n)$ of iid random variables with a regularly varying right tail. For these heavy tailed models and for small values of p, we want to extrapolate beyond the sample, estimating not only the EVI, but also an extreme quantile χ_{1-p}, i.e., a value such that $F(\chi_{1-p}) = 1 - p$, or equivalently,

$$\chi_{1-p} := F^{\leftarrow}(1-p) = U(1/p), \quad p = p_n \to 0, \quad \text{as } n \to \infty,$$

with the notation $F^{\leftarrow}(y) := \inf\{x : F(x) \geq y\}$ for the generalized inverse function of F and $U(t) := F^{\leftarrow}(1 - 1/t)$, $t \geq 1$, for the reciprocal quantile function. In this study, only the classical semi-parametric extreme quantile estimator for heavy right-tails, proposed by [17] will be considered. It is defined as:

$$Q_{p,\hat{\xi}_n^{\bullet}}(k) = X_{n-k:n}\, r_n^{\hat{\xi}_n^{\bullet}(k)}, \quad 1 \leq k < n, \quad r_n \equiv r_n(k;p) = \frac{k}{np}, \tag{7}$$

where $X_{n-k:n}$ is the $(k+1)$-th upper order statistic and $\hat{\xi}_n^{\bullet}$ can be any consistent estimator of the EVI, ξ. The use of other extreme quantile estimators and their comparison is a work in progress.

3 An Overview of Time Series Modelling

A time series is a sequence of observations indexed by time, X_t, usually in equally spaced intervals and correlated. Time series analysis significantly impacts various fields, including economics, finance, medicine, engineering, and environmental studies. By analysing time series, researchers can identify trends, patterns, and anomalies that provide insights into the underlying processes. Using EVT in time series is essential because it allows to identify and model the underlying dependence structure of the data. The search for the best model that describes the stochastic behaviour of a time series is done through statistical techniques. It seeks to explain and describe the variability of data using deterministic functions over time, considering not only its past but also other random variables that may influence the phenomenon under study. That model should be able to capture the time series dynamics in order to be used in the analysis of the structure of the process or for obtaining predictions.

Two of the most widely used time series models are the ARIMA (AutoRegressive Integrated Moving Average) and Exponential Smoothing. ARIMA models are based on linear regression models and use past observations to predict future values. ARIMA models consist of three components: autoregression (AR), differencing (I), and moving average (MA). The autoregression component uses past values of the time series to predict future values. The differencing component is used to remove trends in the data, and the moving average component is used to remove any remaining noise in the data. Exponential smoothing refers to a set of methods that can be used to model and obtain forecasts. This is a versatile approach that continually updates a forecast emphasising the most recent experience. That is, recent observations are given more weight than older observations. Exponential smoothing methods (EXPOS[1]) stand out due to their versatility in the wide choice of models that they include. The widespread dissemination makes them the most widely used method of modelling and forecasting time series. In short, exponential smoothing models take into account both trend and seasonality patterns in the data, while ARIMA models are designed to explain the autocorrelations present in the data [19].

Overall, time series analysis plays a critical key in understanding and predicting data in various fields. Researchers can make informed decisions based on past and current data by using models such as ARIMA and Exponential Smoothing.

4 Application to Environmental Data

Studying and modelling river flow discharge rates is required for river management, including water resources planning, pollution prevention, and flood control. This study considers the daily mean flow discharge rate values (m^3/s) of two rivers in Portugal, Tejo and Guadiana. The time horizon for Tejo and Guadiana are January 1974 until June 2022 and from January 2002 until June 2022, respectively.[2]

Now we are interested in applying EVT to study the behaviour of the maximum values of the time series. As stated before, EVT was designed under iid or weak dependent conditions. Then, taking into account the max-stability property of maxima, the maximum of each month is taken. In Fig. 1 is plotted the maximum value observed in each month. Note that, for each river, there are some months where very high values of river flow discharge are registered.

As an initial descriptive analysis, some descriptive measures for the data set are presented in Table 1.

[1] From [18].

[2] Download from "http://snirh.apambiente.pt at 14/07/2022."

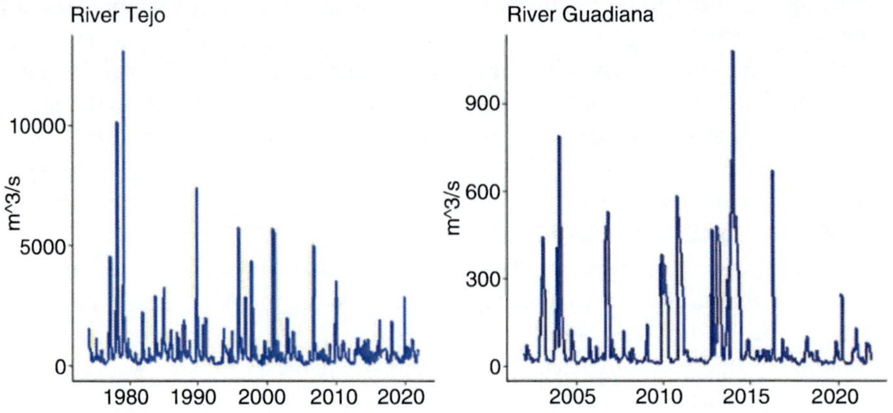

Fig. 1 Monthly flow discharge for Tejo and Guadiana

Table 1 Descriptive statistics measures for both rivers

River	n	min	max	Skewness	Kurtosis
Tejo	576	0.99	13102.55	5.7495	47.373
Guadiana	240	3.35	1080.1	3.2483	12.38033

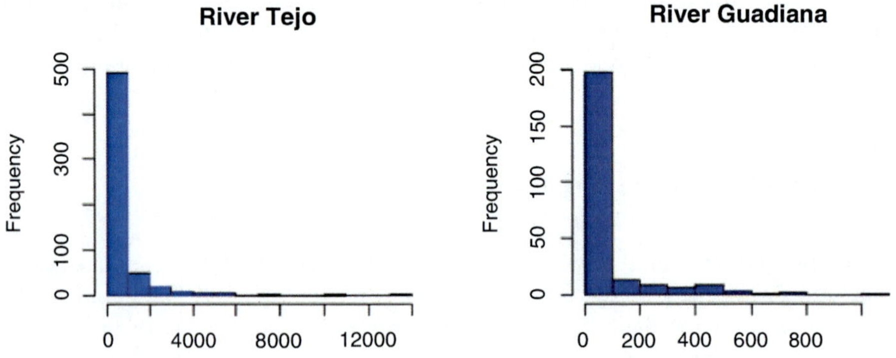

Fig. 2 The histograms for data

Table 2 Accuracy measures for ets() and arima()

	Tejo			Guadiana		
Functions	RMSE	MAE	MAPE	RMSE	MAE	MAPE
ets()	869.76	432.96	98.93	115.34	63.68	145.80
arima()	922.19	426.81	195.99	123.58	67.28	176.53

The data show a skewed pattern and a high kurtosis, denoting heavy right tails. This is also supported by the histograms of both time series shown in Fig. 2.

From now on, all analyses will be performed using the series of monthly maximum values (Fig. 1). The ARIMA and the Exponential Smoothing models were fitted, and some accuracy measures are obtained and presented in Table 2.

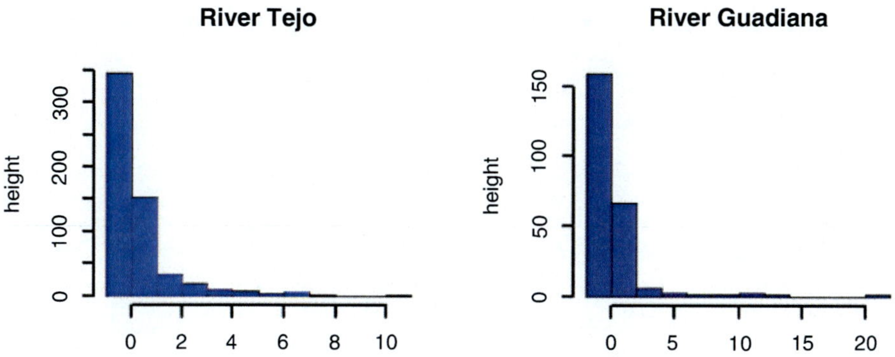

Fig. 3 The histograms for residuals

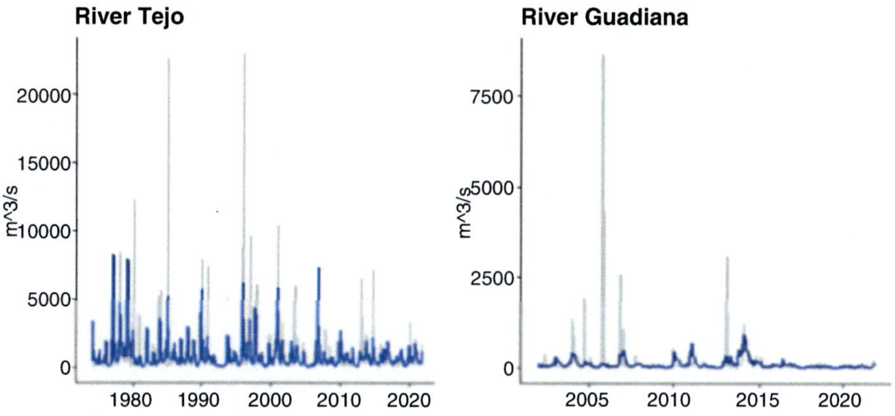

Fig. 4 Time series and the time series reconstructed (grey)

According to this table, Exponential Smoothing is the best model because it has the lowest values of the accuracy measures. The Exponential Smoothing model is fitted to our data, and the residuals are extracted. The EV distribution is fitted to the residuals of the best model. A graphical representation of these residuals in Fig. 3 also revealed a heavier tail. An EV distribution is fitted to the residuals of the model, and based on the EV parameters estimates, a 'reconstructed' time series is obtained, as seen in Fig. 4. Both graphs show that the reconstructed time series captures the extreme values in the original time series.

The main objective of this study is to introduce a methodology that combines EVT and time series analysis to improve the estimation of extremely high quantiles. This is crucial because such quantiles are associated with serious risks, and their estimation is difficult due to the scarcity or absence of observed data. To illustrate the estimation of the key parameter in EVT, ξ, using the estimators described in Sect. 2.2, we present Fig. 5. This figure shows the sample paths of estimates plotted versus k, the number or upper order statistics. As it is well known from EVT

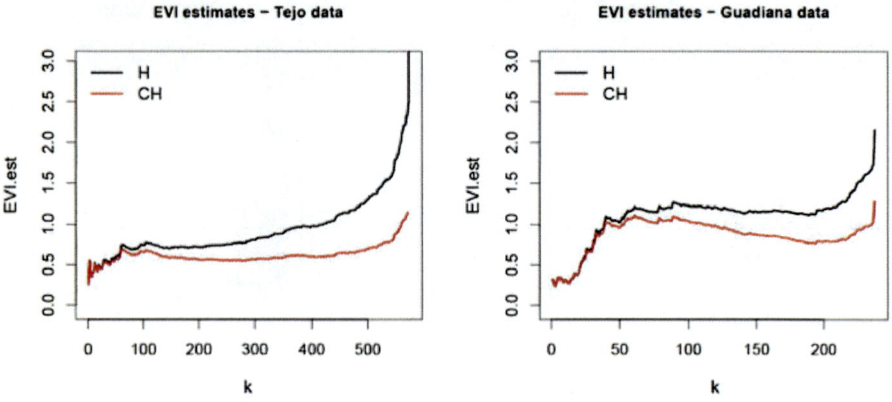

Fig. 5 Sample paths of the EVI-estimates considered

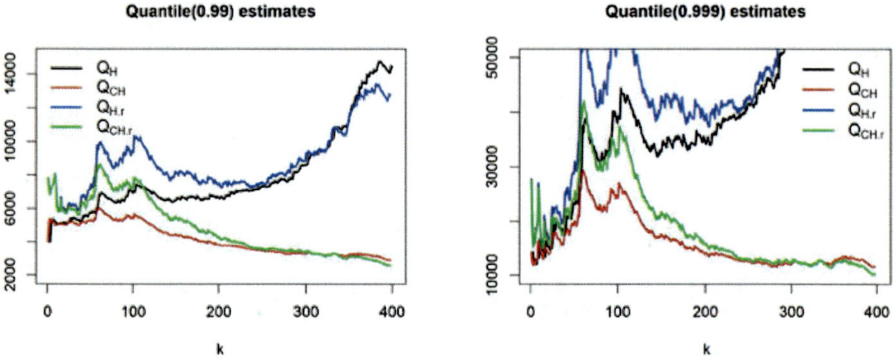

Fig. 6 Sample path of the quantile estimates for Tejo river, based on the original time series and on the reconstructed associated. These are denoted with subscript ".r"

theory, CH estimates have a more stable path. The quantile estimates are calculated when considering the initial time series and the reconstructed time series. For both shape parameter estimators, the sample path of the quantile estimates reveals an underestimation relatively to the sample path obtained over the reconstructed time series, as seen in Figs. 6 and 7.

From these sample paths, we see that the quantiles estimates show some discrepancies, mainly between those based on the H EVI estimator and the CH EVI estimator. This estimator produces more stable paths, and when k increases, the estimates obtained with the original data and the "reconstructed series" show very similar values. So it is now advisable to compare other extreme quantile estimators jointly with computational procedures to choose the "best" value of k (under some criterion) to obtain the quantile estimate.

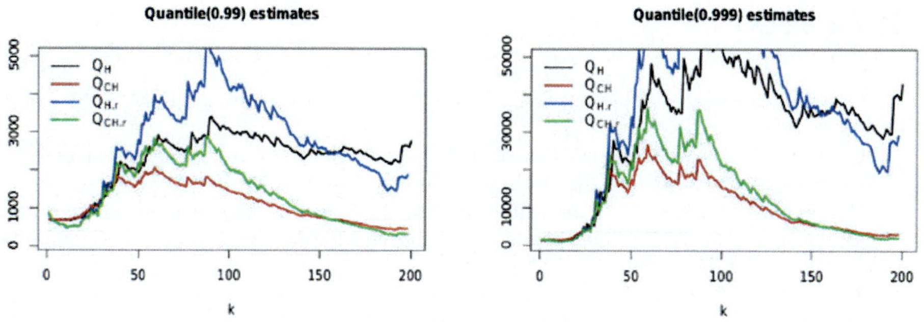

Fig. 7 Sample path of the quantile estimates for Guadiana river, based on the original time series and on the reconstructed associated for river Guadiana. These are denoted with subscript ".r"

5 Concluding Remarks and Future Work

EVT is an important area of statistics that deals with the analysis of rare events or extreme values in a data set. Time series analysis can join EVT modelling extreme events in various fields, such as finance, engineering, and environmental science. In finance, for instance, extreme value theory is used to model and forecast tail events in financial time series, such as stock market crashes and currency crises. In recent literature, there have been several attempts to improve the quality of extreme quantile estimators. It is known that the estimates produced depend on the estimation of EVT key parameters as well as on the quantile estimators themselves. Any of those types of estimators remain topics still under investigation. When trying to estimate very extreme quantiles, we are faced with a lack of observed values. Procedures of modelling and forecasting in time series are important tools to estimate extreme values capturing the characteristics of the series observed. This work joined procedures in both areas to improve the estimation results. Resampling methods associated with the use of algorithms for choosing the number of ordinal statistics to be used in the quantile estimation have revealed as promised procedures. The work in progress considers to use those EVT methodologies jointly with time series to model and forecast extreme quantiles.

Acknowledgments Manuela Neves and Clara Cordeiro are partially financed by national funds through FCT – Fundação para a Ciência e a Tecnologia under the project UIDB/00006/2020. Dora Prata Gomes is financed by national funds through the FCT – Fundação para a Ciência e a Tecnologia, I.P., under the scope of the projects UIDB/00297/2020 and UIDP/00297/2020 (Center for Mathematics and Applications).

References

1. Chavez-Demoulin, V., Davison, A.C.: Modelling time series extremes. REVSTAT Stat. J. **10**(1), 109–133 (2012)
2. Rydell, S.: The use of extreme value theory and time series analysis to estimate risk measures for extreme events. Master's thesis, Umeå University, Umeå (2013)
3. Neves, M.M., Cordeiro, C.: Modelling (and forecasting) extremes in time series: a naive approach. In: Atas do XXXIII Congresso da Sociedade Portuguesa de Estatística, Lisboa, 18–21 Outubro 2017, pp. 189–202 (2020)
4. Yozgatlıgil, C., Türkeş, M.: Extreme value analysis and forecasting of maximum precipitation amounts in the western Black Sea subregion of Turkey. Int. J. Climatol. **38**(15), 5447–5458 (2018)
5. R Core Team. R: A language and environment for statistical computing. R Found. for Statistical Computing, Vienna (2022). https://www.R-project.org/
6. Gumbel, E.J.: Statistics of Extremes. Columbia University Press, New York (1958, 2004)
7. Fréchet, M.: Sur la loi de probabilité de l'écart maximum. Ann. Soc. Polon. Math. (Cracovie) **6**, 93–116 (1927)
8. Fisher, R.A., Tippett, L.H.C.: Limiting forms of the frequency distributions of the largest or smallest member of a sample. Proc. Camb. Philos. Soc. **24**, 180–190 (1928)
9. Gumbel, E.J.: Les valeurs extrêmes des distributions statistiques. Ann. Inst. Henri Poincaré, **5**(2), 115–158 (1935)
10. von Mises, R.: La distribution de la plus grande de n valeurs. American Mathematical Society, Reprinted in Selected Papers Volumen II, Providence (1954), pp. 271–294 (1936)
11. Gnedenko, B.V.: Sur la distribution limite d'une série aléatoire. Ann. Math. **44**, 423–453 (1943)
12. de Haan, L.: On Regular Variation and Its Applications to the Weak Convergence of Sample Extremes. Mathematical Centre Tract 32. D. Reidel, Amsterdam, Dordrecht (1970)
13. Hill, B.M.: A simple general approach to inference about the tail of a distribution. Ann. Stat. **3**, 1163–1174 (1975)
14. Caeiro, F., Gomes, M.I., Pestana, D.D.: Direct reduction of bias of the classical Hill estimator. Revstat **3**, 111–136 (2005)
15. Fraga, A., Gomes, M.I., Laurens, de H.: A new class of semi-parametric estimators of the second order parameter. Port. Math. **60**(2), 193–214 (2003)
16. Gomes, M.I., Martins, M.J.: "Asymptotically unbiased" estimators of the tail index based on external estimation of the second order parameter. Extremes **5**, 5–31 (2002)
17. Weissman, I.: Estimation of parameters and large quantiles based on the k largest observations. J. Am. Stat. Assoc. **73**, 812–815 (1978)
18. DeLurgio, S.A.: Forescasting Principles and Applications. McGraw-Hill International Editions, Boston (1998)
19. Hyndman, R.J., Athanasopoulos, G.: Forecasting: Principles and Practice, 2nd edn. OTexts, Melbourne (2018). Accessed 4th July 2022. http://OTexts.com/fpp2

Factors Associated with Powerful Hurricanes in the Atlantic

Florence George, Sneh Gulati, Anu Simon, and B. M. Golam Kibria

Abstract South Florida is not just a tourist and emigration magnet but also a magnet for hurricanes. Notwithstanding the hurricanes, Covid-19 has made the state more popular than ever leading to even more people moving to the state. This increased migration is leading to more development along the coast and a greater exposure to hurricanes. Add to this is the belief that hurricanes are becoming more powerful due to increased sea surface temperatures and one wonders if South Florida is an environmental disaster waiting to happen. In this paper we investigate if powerful hurricanes are increasing and what factors influence the intensity of hurricanes. To do so, we use a logistic regression model for predicting hurricane intensity. We conclude that mid-level humidity is the most important factor affecting hurricane intensity.

Keywords Atlantic hurricane · Extreme events · Logistic regression

F. George · S. Gulati (✉) · B. M. G. Kibria
Florida International University, Mathematics & Statistics, Miami, FL, USA
e-mail: fgeorge@fiu.edu; gulati@fiu.edu; kibriag@fiu.edu

A. Simon
NOAA, Washington, DC, USA
e-mail: anu.simon@noaa.gov

© The Author(s), under exclusive license to Springer Nature Switzerland AG 2023
C. P. Kitsos et al. (eds.), *Statistical Modelling and Risk Analysis*,
Springer Proceedings in Mathematics & Statistics 430,
https://doi.org/10.1007/978-3-031-39864-3_7

1 Introduction

Being from Miami, you're used to the fact that your home is a vacation spot. But that's what makes Miami one of the best places in the world. We're so rich in different cultures, being so close to Haiti, Cuba, the Dominican Republic, and Puerto Rico, and then you've got people who travel from all over the world just to come to visit. Flo Rida.

There are so many colorful characters in Florida. There's a lot of money, development – not all of it good and corruption. John Grisham

I love Florida. I love the beach. I love the sound of the crashing surfers against the rocks. Emo Philips.

These are just three quotes from a multitude of quotes about Florida found at the website [1]. The Sunshine State, as Florida is commonly known, has always had a record number of people moving to it. According to the Florida Office of Economic Research on April 1, 2021, the population of Florida was estimated to be 21,898,945, a gain of 348,338 residents (1.6%) since the 2020 Census. During the decades of the 1980s, Florida grew by 32.7%; the 1990s by 23.5%; the 2000s by 17.6%; and the 2010s by 14.6% (see [2]). However, we are also the state that is vulnerable to hurricanes and to sea-level rise. With emigration to Florida on the rise, more people are likely to be vulnerable to the devastating effects of hurricanes which seem to be getting bigger and more powerful every year.

The NOAA National Centers for Environmental Information (NCEI) tracks natural disasters and show in a recent report that the U.S. has sustained 323 weather and climate disasters since 1980 where overall damages/costs reached or exceeded $1 billion (U.S. Billion-Dollar Weather and Climate Disasters [3]. In this report, they also examined the distribution of damage from U.S. Billion-dollar disaster events from 1980 to 2021 and show that these are dominated by tropical cyclone losses. Tropical cyclones caused the most damage ($1157.1 billion, adjusted for Consumer Price Index, henceforth referred to as CPI-adjusted, and have the highest average event cost ($20.3 billion per event, CPI-adjusted).

On March 24, 2021, Science Brief [4] published a report discussing the increase in higher intensity hurricanes. According to the article, we are seeing more intense hurricanes worldwide [5]. The results/ hypotheses in the article are based on an examination of more than 90 peer reviewed papers. The article cites [6] to conclude that the proportion of category 3–5 cyclone occurrence has grown by around 5% per decade since 1979. Data also indicate that the likelihood of rapid intensification (defined as an increase in intensity of the tropical storm by at least 18 m/s in 24 hours) has increased [7]. Figures 1 and 2 show the cost (both in dollars and the number of lives lost) incurred from some of the costliest storms to make landfall in the United States. Given the high costs of hurricanes, perhaps governments could enact policy changes in development along the coast and enforce stronger building codes that could serve to avoid the catastrophic damage caused by powerful hurricanes. However, before any changes can be proposed, it is imperative that we investigate the risk from such events and quantify it.

Hurricanes start out as tropical waves in the ocean and are fueled by warm waters and low wind shear to become full -fledged storms. Some recent research

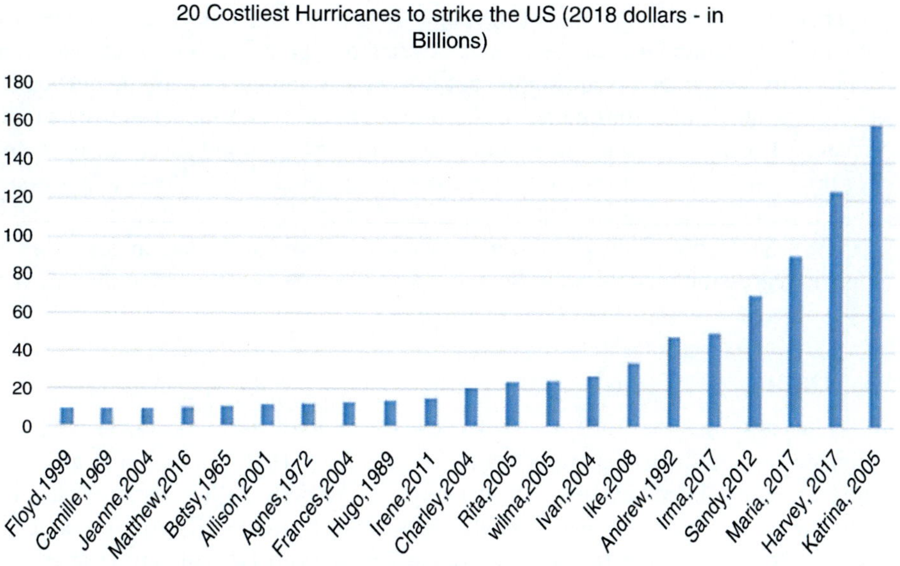

Fig. 1 Twenty costliest hurricanes in US [3]

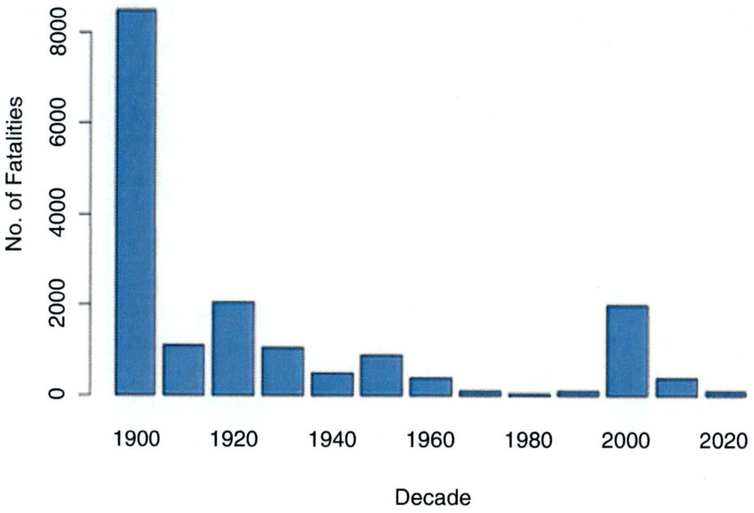

Fig. 2 Hurricane fatalities in US by decades [8]

evaluated the effects of tropical sea surface temperature and vertical wind shear on hurricane development [9]; and the effects of Humidity on Major Hurricanes [10]. For prediction of hurricane intensity, readers may refer to [11–13] among others. In this paper, we investigate if hurricanes are becoming more powerful and what factors have a significant association with hurricane intensity.

Thus, the twofold objective of the paper is to use a logistic regression model in order to (1) examine if it can be predicted whether a hurricane will be extreme, and (2) identify which factors are significant in the development of extreme hurricanes. The covariates to be considered in the development of the model are recent/past (recent 50 years vs prior to that), sea surface temperature, wind shear and humidity. For predicting the occurrence of intense hurricanes or storms by logistic regression model, we refer our readers to [14–16] among others.

The organization of the paper is as follows: We describe the data in Sect. 2, fitted logistic regression models are in Sects. 3 and 4 presents some concluding remarks.

2 Data

This study examined Atlantic based hurricanes from 1940 to 2022, a total of 137 storms ranging from category 1 to category 5. We define a hurricane of category 3 or higher as "Extreme" while below 3 as "Non-Extreme". The purpose of this study is to examine the factors that are associated with Extreme Hurricanes. The variables of interest are Midlevel Humidity (MLH), Sea surface Temperature (SST), Previous month Sea Surface Temperature (PSST) and Wind Shear Atlantic (WSA). The SST values (in deg. C) are area-averaged values in the Main Development Area (MDR) region, which is the area in the Atlantic 10 to 20 N, 85 W to 15 W. Mean Sea level pressure (in hPa, mb) is the average atmospheric pressure at sea level in the Caribbean Sea (9 to 22 N, 89 to 60 W. The vertically averaged (mass) specific humidity (kg) is the proxy for mid-level moisture in Tropical North Atlantic 5.5 N to 23.5 and 15 W to 57 W. The Atlantic Vertical Wind Shear is defined as the difference between the 200- and 850-hPa zonal wind fields.

Preliminary studies of the data show no notable differences in WSA, between the extreme and non-extreme hurricane months. The box plots in Figs. 3, 4 and 5 respectively show some differences in Humidity, Sea Surface Temperature and Previous month temperatures between the extreme and non-extreme hurricane months. Out of the 137 hurricanes during the period 1940–2022, 68 happened in the 1940–1980 period while the remaining 69 happened in the 1980–2022 period, which shows somewhat equal distribution of the number of hurricanes during these two time blocks. However, we will be using all these variables in the logistic regression model in Sect. 3.

As mentioned earlier, out of the 137 hurricanes during the period 1940–2022, 68 happened in the 1940–1980 period while the remaining 69 happened in 1980–2022. There were 22 out of 68 (32.2%) extreme hurricanes in 1940–1980 and 13 out of 69 (18.8%) extreme hurricanes in 1980–2022 period. In other words, out of the total 35 extreme hurricanes in 1940–2022 period, 62.9% happened in 1940–1980 while only 37.1 occurred in recent decades. Even though there is no significant difference between these proportions (p-value = 0.10), the comparatively low proportion of extreme hurricanes in recent decades compared to 1940–1980 was interesting to note and opposite to what we expected. These findings can be observed in Figs. 6 and 7.

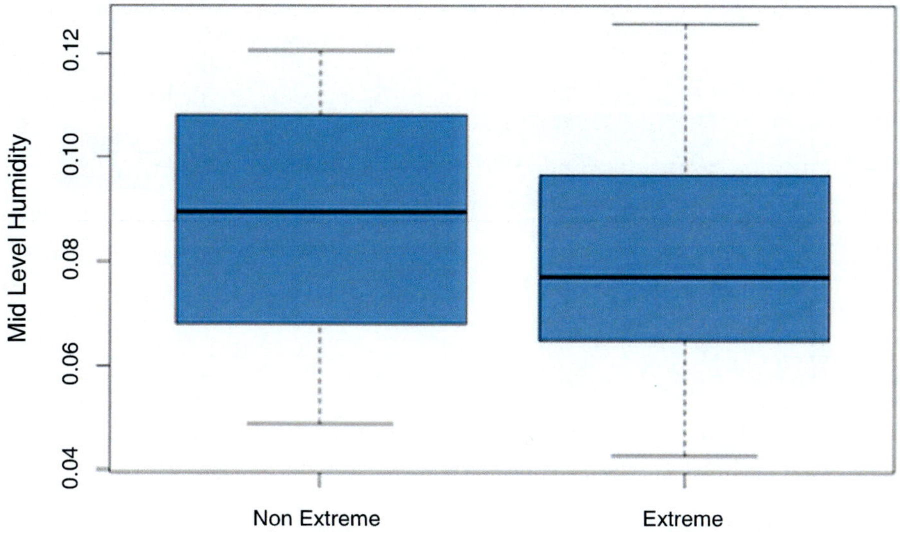

Fig. 3 Midlevel Humidity during the Hurricane Months

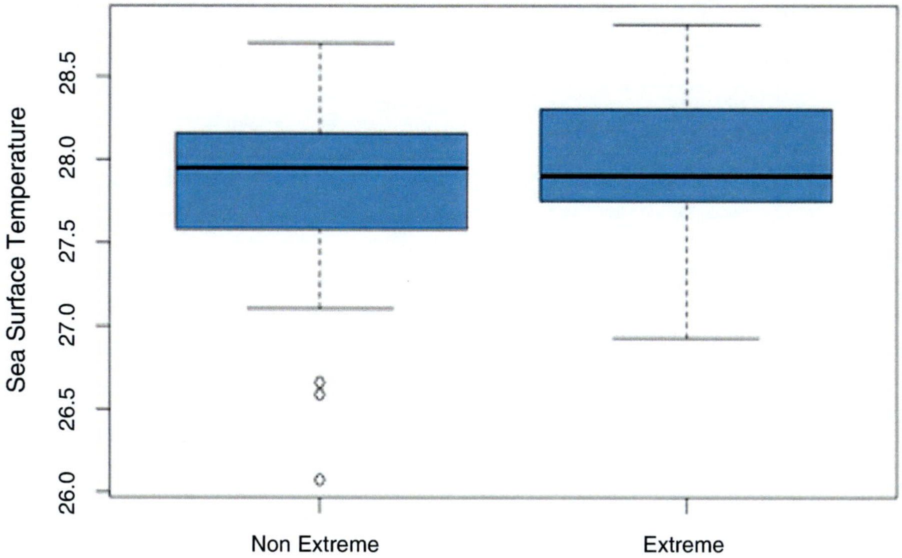

Fig. 4 Sea Surface Temperature during the Hurricane Months

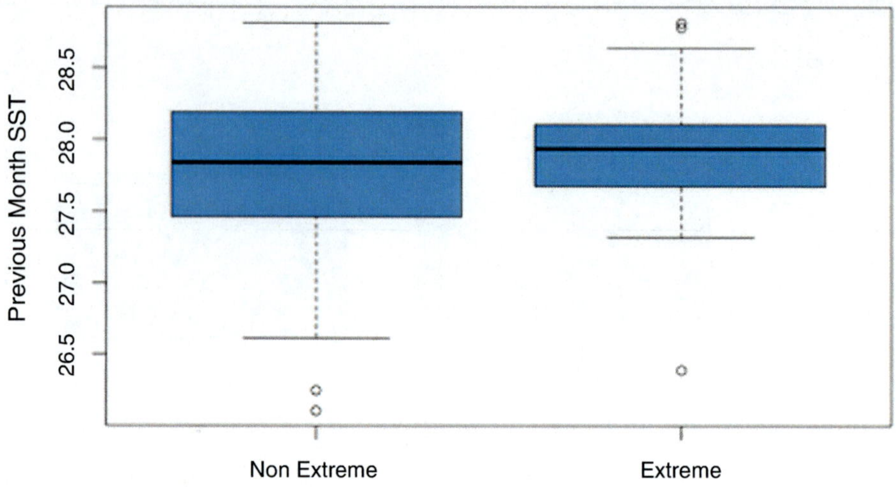

Fig. 5 Sea Surface Temperature in previous Month of Hurricanes

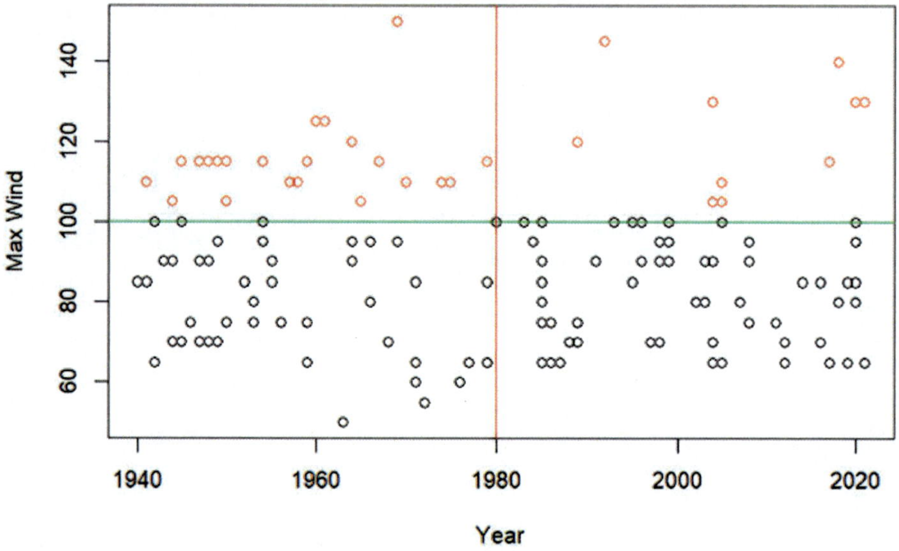

Fig. 6 Hurricane Maximum Wind over time

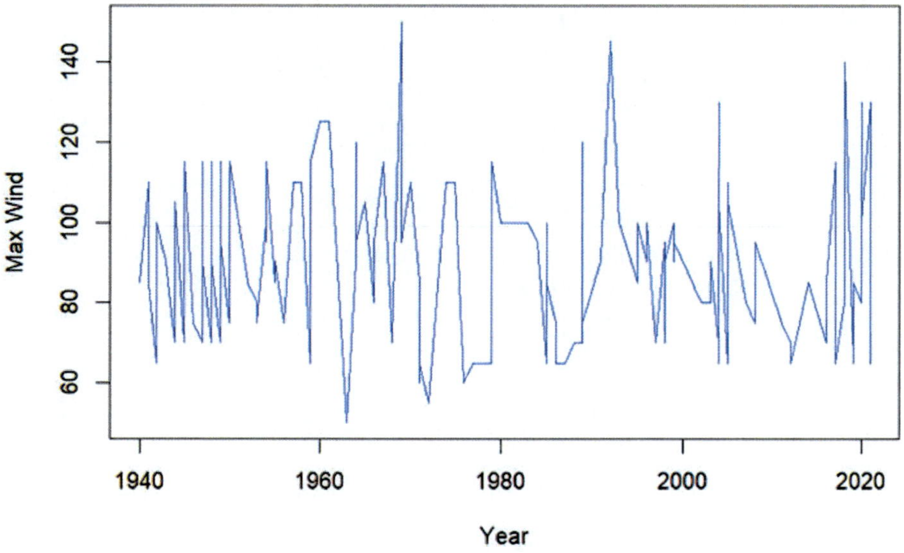

Fig. 7 Time Series Plot of maximum wind

3 Logistic Regression Models, Results and Discussion

Let us assume that y_i is a binary response variable (extreme or non-extreme), then the logistic regression model is defined as Bernoulli distribution $y_i \sim Bernoulli(\pi_i)$ such that

$$\pi_i = \pi(x_i) = \frac{1}{1 + e^{-(x_i'\beta)}} \quad i = 1, 2, ..n \tag{1}$$

where x_i' is a $1xp$ vector of explanatory variables for the ith observation, β is a $px1$ vector of regression coefficients and n is the sample size. The logit transformation of the model in Eq. (1) can be written as

$$\ln\left(\frac{\pi(x)}{1 - \pi(x)}\right) = \beta_0 + \beta_1 x_1 + \cdots + \beta_p x_p \tag{2}$$

For fitting the logistic regression model, we consider the regressors, as defined in the previous Section, and also the following Time variable, to investigate any significant differences between the time groups exist or not.

$$\text{Time} = \begin{cases} 0 \text{ for } 1940 \text{ to } 1979 \\ 1 \text{ for } 1980 \text{ to } 2022 \end{cases}$$

Table 1 Summary of Fitted Logistic Regression Models

Model 1		Estimate	Std. Error	z value	Pr(>|z|)
	(Intercept)	−12.828	15.200	−0.844	0.399
	MLHumidity	−16.994	8.690	−1.956	0.051
	SST	−0.387	0.896	−0.432	0.666
	PSST	0.892	0.665	1.342	0.180
	WSA	−0.036	0.051	−0.702	0.482
	Time (> = 1980)	−0.367	0.402	−0.914	0.361
	AIC = 181.8				
Model 2	(Intercept)	−17.388	10.990	−1.582	0.114
	MLHumidity	−16.587	8.631	−1.922	0.055.
	PSST	0.663	0.396	1.673	0.094
	WSA	−0.019	00.32	−0.583	0.560
	Time (> = 1980)	−0.387	0.400	−0.966	0.334
	AIC = 179.99				
Model 3 (final model)	(Intercept)	−13.961	10.084	−1.384	0.166
	MLHumidity	−17.439	8.570	−2.035	0.042 *
	PSST	0.532	0.362	1.469	0.142
	AIC = 177.1				

We fit the (1) full model keeping all possible regressors, (2) Model with all but SST because SST & PSST have association and (3) Final model chosen using stepwise method. The maximum likelihood method was used to estimate the model parameters. The estimated coefficients associated p-values of variables and AIC in each model are listed in Table 1. It can be observed from the final model in Table 1 that Mid-Level Humidity is the only factor associated with extreme hurricanes. Specifically, as Mid-Level humidity decreases, the odds of extreme hurricane increases.

4 Summary and Concluding Remarks

In this paper we attempted to fit logistic regression models on extreme hurricanes (defined based on maximum wind). We have fitted different models and concluded that Mid-level humidity is associated with extreme hurricanes. Our analysis did not find evidence that hurricane intensity was increasing over time which seemed to contradict previous results [6, 7]. This could be due to the fact that we only examined hurricanes in the mid-Atlantic region and that the sea surface temperature data were averages rather than maxima. We would like to expand this analysis for future studies. This will include looking at all hurricanes and storms in the Atlantic and the Gulf, looking at different time periods and increasing the range of the data. We would also like to explore models on hurricane severity index (HIS), incorporate the intensity of the winds and the size of the area covered by the winds.

Acknowledgement Authors are grateful to the anonymous referees for their valuable comments and suggestions, which certainly improved the quality and presentation of the paper.

References

1. https://thesologlobetrotter.com/florida-quotes-about-florida/
2. http://edr.state.fl.us/Content/population-demographics/reports/econographicnews-2022-v1.pdf
3. https://www.ncei.noaa.gov/access/billions/, https://doi.org/10.25921/stkw-7w73
4. www.sciencebrief.org
5. https://sciencebrief.org/uploads/reviews/ScienceBrief_Review_CYCLONES_Mar2021.pdf
6. Kossin, J.P., Knapp, K.R., Olander, T.L., Velden, C.S.: Global increase in major tropical cyclone exceedance probability over the past four decades. Earth Atmos. Planet. Sci. **117**(22), 11975–11980 (2020)
7. Bhatia, K.T., Vecchi, G.A., Knutson, T.R., et al.: Recent increases in tropical cyclone intensification rates. Nat. Commun. **10**, 635 (2019). https://doi.org/10.1038/s41467-019-08471
8. Czajkowski, J., Simmons, K., Sutter, D.: An analysis of coastal and inland fatalities in landfalling US hurricanes. Nat. Hazards. **59**, 1513–1531 (2011). https://doi.org/10.1007/s11069-011-9849-x
9. Latif, M., Keenlyside, N., Bader, J.: Tropical Sea surface temperature, vertical wind shear, and hurricane development. Geophys. Res. Lett. **34**, L01710 (2007). https://doi.org/10.1029/2006GL027969
10. Pérez-Alarcón, A., Coll-Hidalgo, P., Fernández-Alvarez, J.C., Sorí, R., Nieto, R., Gimeno, L.: Moisture sources for precipitation associated with major hurricanes during 2017 in the North Atlantic basin. J. Geophys. Res. Atmos. **127**, e2021JD035554 (2022)
11. DeMaria, M., Kaplan, J.: An updated statistical hurricane intensity prediction scheme (SHIPS) for the Atlantic and eastern North Pacific basins. Weather Forecast. **14**, 326–337 (1999)
12. DeMaria, M., Mainelli, M., Shay, L.K., Knaff, J.A., Kaplan, J.: Further improvements to the statistical hurricane intensity prediction scheme (SHIPS). Weather Forecast. **20**, 531–543 (2005)
13. Law, K.T., Hobgood, J.S.: A statistical model to forecast short-term Atlantic hurricane intensity. Weather Forecast. **22**, 967–980 (2007)
14. Jing, B., Qian, Z., Zareipour, H., Pei, Y., Wanf, A.: Wind turbine power curve modelling with logistic functions based on quantile regression. Appl. Sci. **11**, 3048 (2021). https://doi.org/10.3390/app11073048
15. Kovacs, J.M., Blanco-Correa, M., Flores-Verdugo, F.: A logistic regression model of hurricane impacts in a mangrove Forest of the Mexican Pacific. J. Coast. Res. **17**(1), 30–37 (2001)
16. Srivastava, N.: A logistic regression model for predicting the occurrence of intense geomagnetic storms. Ann. Geophys. **23**, 2969–2974 (2005)

Reliable Alternative Ways to Manage the Risk of Extreme Events

M. Ivette Gomes, Fernanda Figueiredo, and Lígia Henriques-Rodrigues

Abstract In the field of statistical extreme value theory, risk is generally expressed either by the *value at risk* at a level q (VaR_q), the upper $(1-q)$-quantile of the loss function, or by the *conditional tail expectation* (CTE), defined as $\text{CTE}_q = \mathbb{E}(X|X > \text{VaR}_q)$, $q \in (0, 1)$. We consider heavy-tailed models, i.e. Pareto-type underlying CDFs, with a positive *extreme value index* (EVI), quite common in many areas of application. For these Pareto-type models, the classical EVI-estimators are the Hill (H) estimators, the average of the k log-excesses over a threshold $X_{n-k:n}$. The Hill estimator is crucial for the semi-parametric estimation of both the VaR and the CTE. We now suggest an improvement in the performance of the aforementioned CTE-estimators, through the use of a reliable EVI–estimator based on generalized means and possibly reduced-bias.

Keywords Conditional tail expectation · Generalized means · Heavy-tailed parents · Risk modeling · Semi-parametric estimation

M. I. Gomes (✉)
DEIO, Faculdade de Ciências, Universidade de Lisboa, Lisboa, Portugal

Centro de Estatística e Aplicações da Universidade de Lisboa (CEAUL), Lisboa, Portugal
e-mail: ivette.gomes@ciencias.ulisboa.pt

F. Figueiredo
Centro de Estatística e Aplicações da Universidade de Lisboa (CEAUL), Lisboa, Portugal

Faculdade de Economia, Universidade do Porto, Porto, Portugal

L. Henriques-Rodrigues
Departamento de Matemática e CIMA, Universidade de Évora, Évora, Portugal

© The Author(s), under exclusive license to Springer Nature Switzerland AG 2023
C. P. Kitsos et al. (eds.), *Statistical Modelling and Risk Analysis*,
Springer Proceedings in Mathematics & Statistics 430,
https://doi.org/10.1007/978-3-031-39864-3_8

1 Introduction and Scope of the Paper

In the field of statistical *extreme value theory* (EVT), a great variety of alternative methodologies are available to deal with the management of risks of extreme events, like a big loss that occurs very rarely.

The risk is generally expressed either by the *value at risk* at a level q (VaR_q), the q-quantile of the random variable X, with a *cumulative distribution function* (CDF) $F(x) = \mathbb{P}(X \le x)$, defined as

$$\text{VaR}_q := Q(q), \quad \text{with } Q(q) := \inf\{x \ge 0 : F(x) \ge q\}, \quad q \in (0, 1), \tag{1}$$

or by the *conditional tail expectation* (CTE), defined as

$$\text{CTE}_q \equiv \text{CTE}_q(X) = \mathbb{E}(X|X > Q(q)), \quad q \in (0, 1), \tag{2}$$

with $Q(\cdot)$ defined in (1). The CTE measure is more informative than VaR, as can be seen in [1], among others. Indeed, the VaR has two important limitations as a risk measure: it is only a quantile of the profit-and-loss distribution, and therefore does not give any information about the potential loss beyond the VaR level; and it is not a coherent risk measure because it is not subadditive, and thus, may discourage the diversification of the portfolios. Contrarily to the VaR, the CTE deals with these two limitations, as referred in [2], and is becoming an alternative to the VaR as a risk measure.

Since risks are more dangerous for heavy right-tails, we further assume that the right-tail function is a Pareto-type tail, i.e.

$$1 - F(x) = x^{-1/\xi}\mathbb{L}(x), \quad \xi > 0, \tag{3}$$

where $\mathbb{L}(\cdot)$ is a slowly varying function at infinity, i.e. $\mathbb{L}(tx)/\mathbb{L}(t) \to 0$, as $t \to \infty$, for all $x > 0$. Equivalently, we can say that $1 - F$ is a regularly varying function at infinity, with an index of regular variation equal to $-1/\xi$, i.e. $1 - F \in \mathcal{R}_{-1/\xi}$. Then, and given a random sample $\underline{\mathbf{X}}_n := (X_1, \ldots, X_n)$ from $F(x)$, Gnedenko's extremal types theorem holds for $X_{n:n} := \max(X_1, \ldots, X_n)$ [3], i.e. the limiting CDF of $X_{n:n}$, linearly normalized, is necessarily of the type of the *extreme value* (EV) CDF,

$$\text{EV}_\xi(x) = \exp(-(1 + \xi x)^{-1/\xi}), 1 + \xi x > 0, \text{ with } \xi > 0. \tag{4}$$

The CDF F is then said to belong to the max-domain of attraction of EV_ξ, and we write $F \in \mathcal{D_M}\left(\text{EV}_{\xi>0}\right) =: \mathcal{D_M}^+$. The parameter ξ, which can more generally be any real number, is the so-called *extreme value index* (EVI), the primary parameter of extreme events.

1.1 EVI-estimation of Pareto-Type Models

For the Pareto-type models, in (3), the classical EVI-estimators are the Hill (H) estimators [4],

$$\mathrm{H}_{k,n} \equiv \mathrm{H}(k; \underline{\mathbf{X}}_n) := \frac{1}{k}\sum_{i=1}^{k} \ln X_{n-i+1:n} - \ln X_{n-k:n}, \quad 1 \le k < n. \tag{5}$$

Recently, several reliable EVI-estimators based on *generalized means* (GMs) (see [5–10], and references therein) have been introduced in the literature. Among them, we refer the *mean-of-order-p* (MO_p) EVI-estimators, initially considered almost simultaneously in [11–13] (see also [14]). The MO_p EVI-estimators are associated with the statistics,

$$U_{ik} := \frac{X_{n-i+1:n}}{X_{n-k:n}}, \quad 1 \le i \le k < n,$$

and defined by

$$\mathrm{H}_{k,n,p} := \begin{cases} \left(1 - \left(\frac{1}{k}\sum_{i=1}^{k} U_{ik}^{p}\right)^{-1}\right)/p, & \text{if } p < 1/\xi,\ p \neq 0, \\ \frac{1}{k}\sum_{i=1}^{k} \ln U_{ik} = \mathrm{H}_{k,n}, & \text{if } p = 0. \end{cases} \tag{6}$$

The use of the extra tuning parameter $p \in \mathbb{R}$ and the MO_p methodology can indeed provide a much more adequate EVI-estimation.

1.2 Further Details on the CTE

Assuming that F is continuous, we can rewrite $\mathrm{CTE}_q(X)$, in (2), as

$$\mathbb{C}_q(X) \equiv \mathrm{CTE}_q(X) = \frac{1}{1-q}\int_q^1 Q(s)ds,$$

and a natural estimator of $\mathbb{C}_q(X)$ can then be obtained by

$$\widehat{\mathbb{C}}_{n,q}(\underline{\mathbf{X}}_n) = \frac{1}{1-q}\int_q^1 Q_n(s)ds, \tag{7}$$

where $Q_n(s)$ is the empirical quantile function,

$$Q_n(s) := X_{i:n} \quad \forall\, s \in \left(\tfrac{i-1}{n}, \tfrac{i}{n}\right] \text{ and } i = 1, \ldots, n.$$

The asymptotic behaviour of the estimator $\widehat{\mathbb{C}}_{n,q}(\underline{\mathbf{X}}_n)$, in (7), has been studied in [15], when $\mathbb{E}(X^2) < \infty$, and more generally, for $\mathbb{E}(X) < \infty$, in [16] and [17]. As already stated in [16], the classical moment assumption, $\mathbb{E}(X^2) < \infty$, is quite restrictive. Indeed, assume that F is the Pareto distribution with index $\xi > 0$, i.e. $1 - F(x) = x^{-1/\xi}$ for all $x \geq 1$. Let us focus on the case $\xi < 1$, because when $\xi \geq 1$, then $\text{CTE}_q(X) = +\infty$, for every $q \in (0, 1)$. If $\xi \in (0, 1/2)$, $\mathbb{E}(X^2) < \infty$. When $\xi \in (1/2, 1)$, we have $\mathbb{E}(X^2) = \infty$ but, nevertheless, $\text{CTE}_q(X)$ is well defined and finite since $\mathbb{E}(X) < \infty$. Analogous remarks hold for the Pareto-type tails, in (3). Note that, in the case $\xi = 1/2$, the finiteness of the second moment depends on the slowly varying function $\mathbb{L}(\cdot)$ in (3).

Notice next that $\mathbb{C}_q(X)$ can be rewritten as

$$\begin{aligned}\mathbb{C}_q(X) &= \frac{1}{1-q}\int_q^{1-k/n} Q(s)ds + \frac{1}{1-q}\int_{1-k/n}^{1} Q(s)ds \\ &=: \mathbb{C}_{k,q}^{(1)}(X) + \mathbb{C}_{k,q}^{(2)}(X).\end{aligned}$$

In this spirit, and with $\text{H}_{k,n}$ the Hill estimator in (5), the authors in [16] (see also [18]) introduced the following estimator of the CTE,

$$\begin{aligned}\widetilde{\mathbb{C}}_{k,n,q}(\underline{\mathbf{X}}_n; \text{H}_{k,n}) &= \widetilde{\mathbb{C}}_{k,n,q}^{(1)}(\underline{\mathbf{X}}_n) + \widetilde{\mathbb{C}}_{k,n,q}^{(2)}(\underline{\mathbf{X}}_n; \text{H}_{k,n}) \\ &= \frac{1}{1-q}\int_q^{1-k/n} Q_n(s)ds + \frac{k X_{n-k:n}}{n(1-q)(1-\text{H}_{k,n})},\end{aligned}$$

written in [17] as

$$\begin{aligned}\widetilde{\mathbb{C}}_{k,n,q}(\underline{\mathbf{X}}_n; \text{H}_{k,n}) = \frac{1}{1-q}\sum_{j=1}^{n-k}\left(\left(\tfrac{j}{n} - q\right)_+ - \left(\tfrac{j-1}{n} - q\right)_+\right) X_{j:n} + \\ + \frac{k X_{n-k:n}}{n(1-q)(1-\text{H}_{k,n})},\end{aligned} \tag{8}$$

where $s_+ := \max(s, 0)$. Note that the estimator $\widetilde{\mathbb{C}}_{k,n,q}^{(1)}(\underline{\mathbf{X}}_n)$ is obtained in the lines of (7), through the use of the well-known properties of the empirical quantile function Q_n, whereas $\widetilde{\mathbb{C}}_{k,n,q}^{(2)}(\underline{\mathbf{X}}_n; \text{H}_{k,n})$ is obtained using a Weissman estimator of $Q(\cdot)$, in (1) [19]:

$$\widehat{Q}(1-q) := X_{n-k:n}\left(\frac{k}{nq}\right)^{\text{H}_{k,n}}, \text{ as } q \to 0.$$

1.3 Scope of the Paper

Since $\mathrm{H}_{k,n}$ can be replaced in (8) by any consistent EVI-estimator, we now suggest an improvement in the performance of the aforementioned CTE-estimators, in (8), through the use of the more general and reliable EVI-estimators in (6). After a brief study, in Sect. 2, of the asymptotic properties of

$$\widetilde{\mathbb{C}}_{k,n,p,q}(\underline{\mathbf{X}}_n) \equiv \widetilde{\mathbb{C}}_{k,n,q}(\underline{\mathbf{X}}_n;\, \mathrm{H}_{k,n,p})$$

we provide in Sect. 3, a Monte-Carlo simulation study of those MO_p CTE-estimators. Some concluding remarks will be put forward in Sect. 4.

2 A Few Considerations on the Asymptotic Behaviour of the MO_p CTE-estimators

In order to be able to study the asymptotic behaviour of the CTE-estimator, in (8), or more generally of

$$\begin{aligned}\widetilde{\mathbb{C}}_{k,n,p,q}(\underline{\mathbf{X}}_n) &\equiv \widetilde{\mathbb{C}}_{k,n,q}(\underline{\mathbf{X}}_n;\, \mathrm{H}_{k,n,p}) \\ &= \widetilde{\mathbb{C}}^{(1)}_{k,n,p,q}(\underline{\mathbf{X}}_n) + \widetilde{\mathbb{C}}^{(2)}_{k,n,q}(\underline{\mathbf{X}}_n;\, \mathrm{H}_{k,n,p}) \\ &= \frac{1}{1-q}\sum_{j=1}^{n-k}\left(\left(\tfrac{j}{n}-q\right)_+ - \left(\tfrac{j-1}{n}-q\right)_+\right) X_{j:n} + \\ &\qquad\qquad \frac{k X_{n-k:n}}{n(1-q)(1-\mathrm{H}_{k,n,p})},\end{aligned} \tag{9}$$

with $\mathrm{H}_{k,n,p}$ given in (6), it is sensible to impose a second-order expansion on the tail function $1-F(x)$, in (3), or on the *reciprocal quantile function*

$$U(t) := F^{\leftarrow}(1-1/t),\ t \geq 1,$$

which is of regular variation with index ξ [20], i.e. $U \in \mathcal{R}_\xi$. Here we shall assume that we are working in Hall-Welsh class of models [21], where, as $t \to \infty$ and with $C,\ \xi > 0$, $\rho < 0$ and β non-zero,

$$U(t) = Ct^{\xi}\left(1 + A(t)/\rho + o\left(t^{\rho}\right)\right), \qquad A(t) = \xi\,\beta\,t^{\rho}. \tag{10}$$

The class in (10) is a wide class of models, that contains most of the heavy-tailed parents useful in applications, like the *Fréchet*, the *Generalized Pareto* and the *Student-*t_ν, with ν degrees of freedom.

In the lines of **Theorem 1** in [17], and just as in **Lemma 1** of that same article, we can write

$$\frac{n(1-q)}{\sqrt{k}\,U(n/k)}\left(\widetilde{\mathbb{C}}^{(2)}_{k,n,q}(\underline{\mathbf{X}}_n;\mathrm{H}_{k,n,p})-\mathbb{C}^{(2)}_{k,q}(X)\right)=\sum_{i=1}^{4}T_{n,i}.$$

The main difference regarding asymptotic bias of $\widetilde{\mathbb{C}}_{k,n,p,q}(\underline{\mathbf{X}}_n)$, in (9), lies on the fact that the scale for $A(n/k)$ in $T_{n,3}$ is given by

$$\frac{1-p\xi}{(1-\rho-p\xi)(1-\xi)^2}$$

and

$$b_p(\xi,\rho)\equiv B(p,\xi,\rho)=\frac{1-p\xi}{(1-\rho-p\xi)(1-\xi)^2}+\frac{1}{(1-\xi)(\xi+\rho-1)}.$$

Regarding the variance, cumbersome computations, of the type of the ones in [17], with their $\mathbb{W}_{n,3}$ replaced by $(1-p\xi)\mathbb{W}_{n,3}/\sqrt{1-2p\xi}$, lead us to

$$v_p(\xi)\equiv V(p,\xi)=\frac{2\xi}{2\xi-1}+\frac{\xi^2}{(1-\xi)^2}+\frac{\xi^2(1-p\xi)^2}{(1-\xi)^4(1-2p\xi)}+\frac{2\xi}{1-\xi}.$$

We can thus state, without the need of a proof, the following:

Theorem 1 *Assume that* (10) *holds, with* $\xi\in(0,1)$, $\xi\neq 1/2$, $p<1/(2\xi)$ *and* $p\neq 1$, $k=k_n$ *is an intermediate sequence of integers, i.e.*

$$k=k_n\to\infty,\quad k/n\to 0,\ \textit{as}\ n\to\infty,$$

and

$$\sqrt{k}A(n/k)\to\lambda,\ \textit{finite}.$$

Then

$$\frac{n(1-q)}{\sqrt{k}U(n/k)}\left(\widetilde{\mathbb{C}}_{k,n,p,q}(\underline{\mathbf{X}}_n)-\mathbb{C}_q(X)\right)\underset{n\to\infty}{\overset{d}{\longrightarrow}}\textit{Normal}\left(\lambda\,b_p(\xi,\rho),v_p(\xi)\right),$$

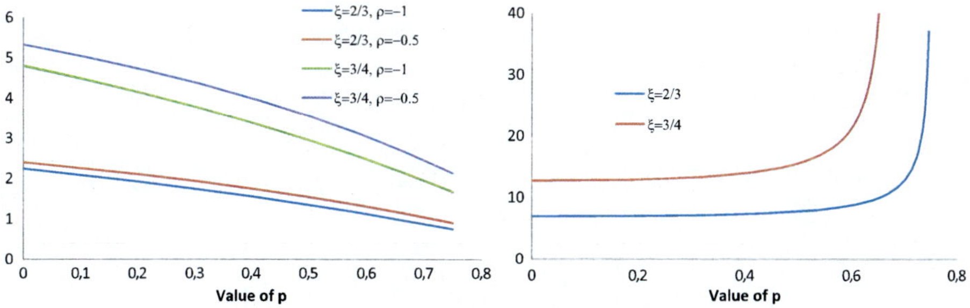

Fig. 1 Asymptotic bias (left) and asymptotic standard deviation indicators (right) of the CTE-estimators for $\xi = 2/3, 3/4$ and $\rho = -0.5, -1$ as a function of p

where

$$b_p(\xi, \rho) = \frac{\xi\rho(1-p)}{(1-\xi)^2(\xi+\rho-1)(1-\rho-p\xi)} \tag{11}$$

and

$$v_p(\xi) = \frac{\xi^4\left[1-2p\xi+p^2(2\xi-1)\right]}{(1-\xi)^4(2\xi-1)(1-2p\xi)}. \tag{12}$$

Figure 1 aims to illustrate the asymptotic behaviour of the MO_p CTE-estimators as function of p, for $\xi \in (1/2, 1)$ and $\rho < 0$. If $\xi < 1/2$, $\mathbb{E}(X^2) < \infty$, and this case has been already studied. For a better visualization of this behaviour, we only represent the asymptotic bias $b_p(\xi, \rho)$ and the asymptotic standard deviation $\sigma_p(\xi) = \sqrt{v_p(\xi)}$ indicators, with $b_p(\xi, \rho)$ and $v_p(\xi)$ respectively given in (11) and (12), for values of $\xi = 2/3,\ 3/4$, $\rho = -0.5,\ -1$ and $p < 1/(2\xi)$. In this illustration we only consider values of p for which we get a positive and finite asymptotic standard deviation. We can observe that the asymptotic bias, $b_p(\xi, \rho)$, is always a decreasing function with p, whereas the asymptotic standard deviation, $\sigma_p(\xi)$, is an increasing function in p. Note that the value $p = 0$ leads to the CTE-estimator in (8), and that it is possible to find a value of p that allows for a significant decrease in the asymptotic bias keeping the asymptotic standard deviation close to the one in (8).These results led us to compare the MO_p CTE-estimators at optimal levels, in Sect. 3.

As expected, the asymptotic bias and standard deviation indicators are significantly large as ξ approaches 1, and an adequate choice of p allows to obtain a MO_p CTE-estimator with an interesting performance, comparatively to the classical CTE-estimator.

Remark 1 Further note that when $p = 1$ the estimator $\widetilde{\mathbb{C}}_{k,n,p,q}(\underline{\mathbf{X}}_n)$ in (9) is given by the expression

$$\widetilde{\mathbb{C}}_{k,n,p,q}(\underline{\mathbf{X}}_n) = \frac{1}{1-q}\left[\sum_{j=1}^{n-k}\left(\left(\tfrac{j}{n}-q\right)_+ - \left(\tfrac{j-1}{n}-q\right)_+\right)X_{j:n} + \frac{1}{n}\sum_{i=1}^{k}X_{n-i+1:n}\right]$$
$$= \frac{1}{[n(1-q)]}\left[\sum_{j=[qn]+1}^{n-k} X_{j:n} + \sum_{i=1}^{k} X_{n-i+1:n}\right]$$
$$= \frac{1}{[n(1-q)]}\sum_{j=[qn]+1}^{n} X_{j:n}, \qquad (13)$$

which is the average of the $[n(1-q)]$ upper order statistics, and where $[x]$ denotes, as usual, the integer part of x.

Remark 2 When $\xi = 1/2$ and $p \le 1/\xi$, or when $p = 1$ and $\xi < 1$, we can still guarantee the consistency of the CTE–estimator $\widetilde{\mathbb{C}}_{k,n,p,q}(\underline{\mathbf{X}}_n)$. But we are then under non-regular frameworks of the type of the one considered in [22–24] for the MO_p EVI-estimation, and $p \ge 1/(2\xi)$.

3 A Monte-Carlo Simulation Study

In this section, we perform a small scale Monte-Carlo simulation study to assess the performance of the new class of CTE-estimators introduced in (9) for values of $\xi \in [0.5, 1)$ and where $\mathbb{E}(X^2) = \infty$. A few values of p were selected previously to enhance the performance of the new estimators. The value $p = 0$ was also included as it provides the CTE-estimator in (8) and we have worked with $p = \{0, 0.5, 0.75, 1, 1.25\}$. The Pareto-type models considered were:

1. The Burr(ξ, ρ) model, with distribution function $F(x) = 1 - (1 + x^{\rho/\xi})^{1/\rho}$, for $x > 0$, with $\xi = \{2/3, 0.5\}$ and $\rho = \{-0.5, -1\}$;
2. The Generalized Pareto model, GP(ξ) with, distribution function $F(x) = 1 - (1 + \xi x)^{-1/\xi} = 1 + \ln \mathrm{EV}_\xi(x)$, with EV_ξ given in (4), for values of $x \ge -1/\xi$ and with $\xi = \{3/4, 2/3, 0.5\}$.

For each model, 1000 samples of sizes $n = \{100, 200, 500, 1000, 2000\}$ were generated and associated CTE-estimates were computed, $\widetilde{\mathbb{C}}_{k,n,p,q}(\underline{\mathbf{X}}_{n,i})$, $k = 1, \ldots, n-1$, $i = 1, 2, \ldots, 1000$. The behaviour of those estimates, as function of k, for $n = 500$, is presented in Figs. 2, 3, and 4. As in [17] we present the simulated median values $(\mathrm{Med}\{\widetilde{\mathbb{C}}_{k,n,p,q}(\underline{\mathbf{X}}_{n,i}), i = 1, \ldots, 1000\}, 1 \le k \le n-1)$, at the left, and the simulated median square errors $(\mathrm{Med}\{(\widetilde{\mathbb{C}}_{k,n,p,q}(\underline{\mathbf{X}}_{n,i}) - \mathbb{C}_q(X))^2, i = 1, \ldots, 1000\}, 1 \le k \le n-1)$, at the right. The choice of q plays a crucial role in

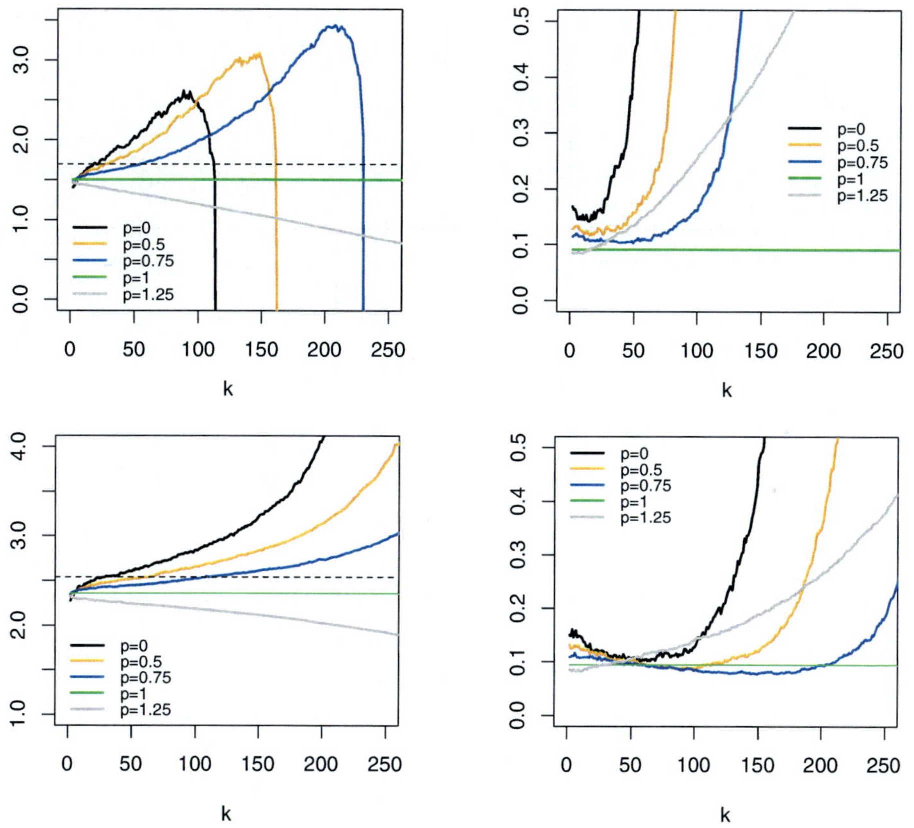

Fig. 2 Simulated median values (left) and simulated median square errors (right) of the CTE-estimators under study from samples of size $n = 500$, for a Burr model with $\xi = 2/3$, $\rho = -0.5$ (top) and $\rho = -1$ (bottom), $q = 0.05$ and $p = \{0, 0.5, 0.75, 1, 1.25\}$

the estimation of the CTE and we have chosen the value $q = 0.05$ to illustrate the finite sample properties of the new estimators. The theoretical CTEs, represented by the horizontal dashed lines in the left panel of the figures, were computed using the package VaRES [25, 26] available in the R software [27].

The following conclusions can be drawn from the simulation results presented in Figs. 2, 3, and 4:

- The simulated median values of the CTE-estimator in (8) are above the true value of the CTE for a wide region of k values. The same happens with the new class of estimators in (9), whenever $p < 1$, but for a smaller region of k values that depends on the choice of p. The reverse happens for $p \geq 1$;
- The simulated median paths are more unstable for values of ξ closer to 1;
- The simulated median square errors exhibit the usual shape associated with the simulated mean square errors. For larger values of ξ the CTE-estimators are very sensitive to the choice of k;

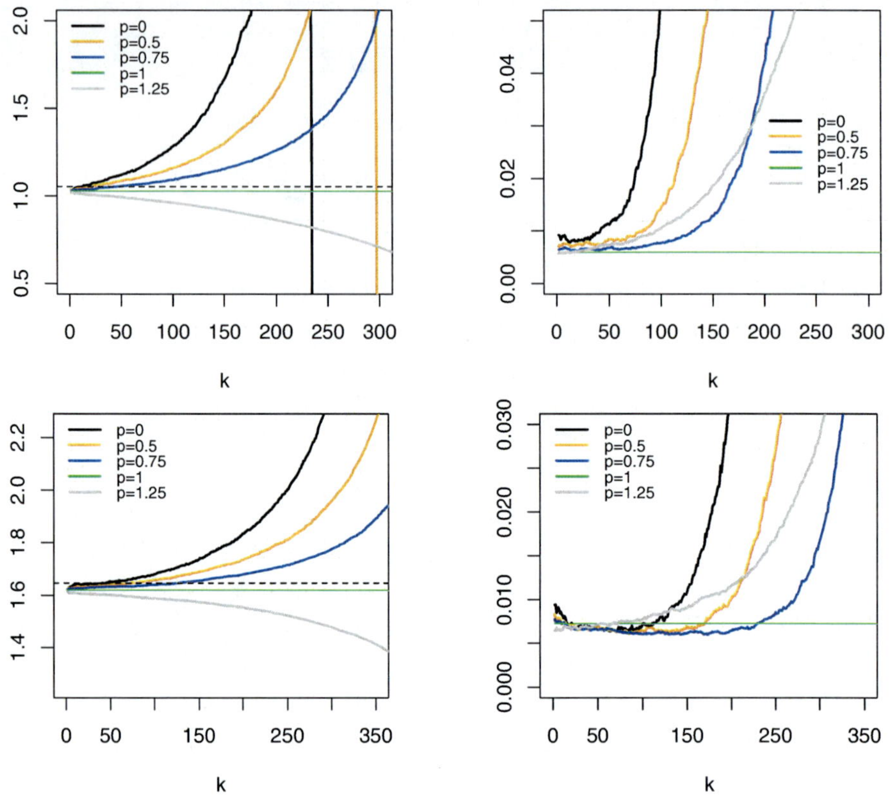

Fig. 3 Simulated median values (left) and simulated median square errors (right) of the CTE-estimators under study from samples of size $n = 500$ for a Burr model with $\xi = 0.5$, $\rho = -0.5$ (top) and $\rho = -1$ (bottom), $q = 0.05$ and $p = \{0, 0.5, 0.75, 1, 1.25\}$

- In the Burr model, for a fixed ξ, larger values of ρ lead to more unstable simulated median paths and larger simulated median square errors.
- It is always possible to find a value of p, positive, such that the new class of CTE-estimators in (9) outperforms in terms of simulated median values and simulated median square errors the CTE-estimator in (8).

In Tables 1 and 2 we present the simulated median values at the simulated optimal levels, the levels that minimize the simulated median square error, and in Tables 3 and 4 we present the simulated median square errors at optimal levels. For each sample size, the simulated median value corresponding to the smallest absolute median-bias, and the smallest simulated median square error, are written in bold. The smallest absolute median-bias depends upon the selected parent, the sample size and the values of (ξ, ρ). For the GP parent and for the ξ-values under study the choice of $p = 0.75$ seems to be adequate. For the Burr models considered, the values $p = 0.5$ and $p = 0.75$ provide the best results. Note that for the Burr

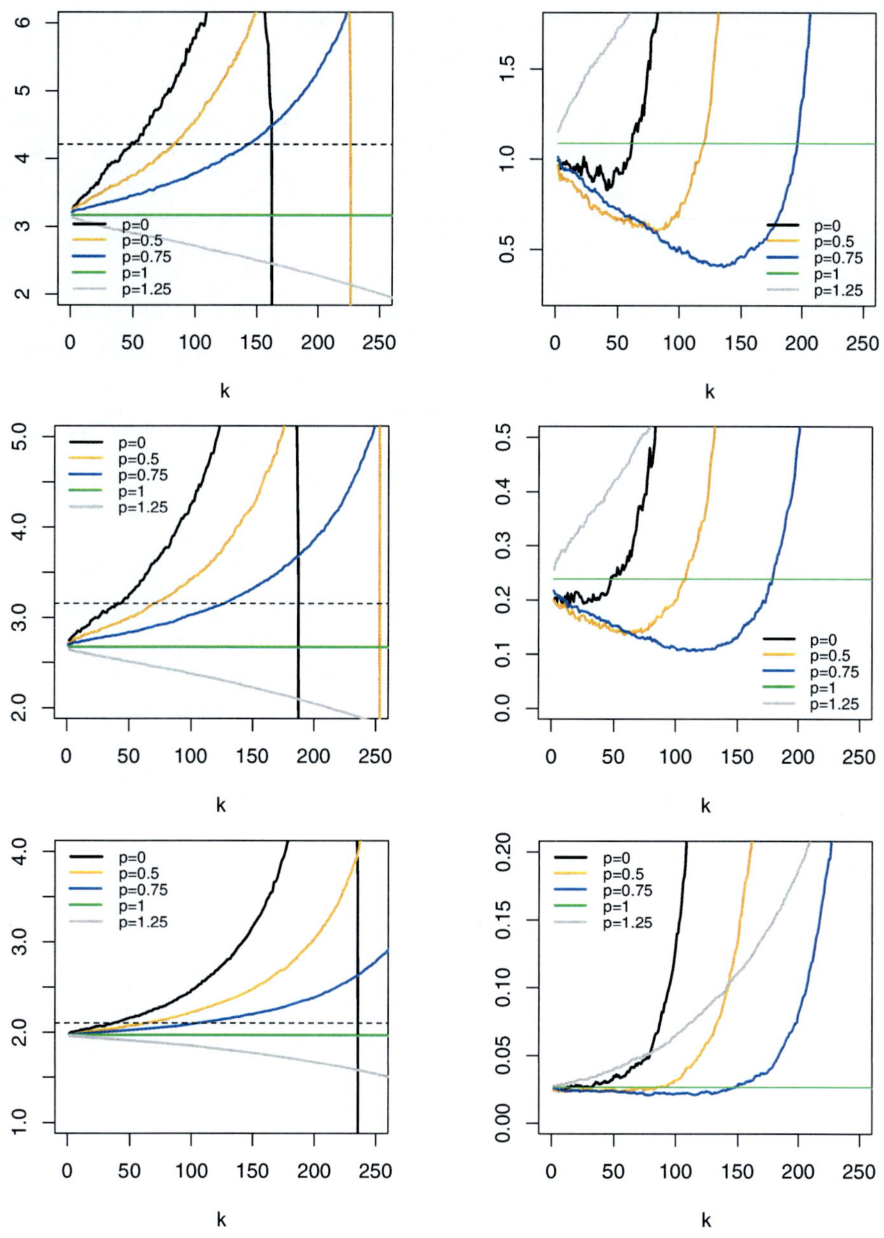

Fig. 4 Simulated median values (left) and simulated median square errors (right) of the estimators under study from samples of size $n = 500$, for a GP model with $\xi = 3/4$ (top), $\xi = 2/3$ (middle) and $\xi = 0.5$ (bottom), $q = 0.05$ and $p = \{0, 0.5, 0.75, 1, 1.25\}$

Table 1 Simulated medians at the simulated optimal level for Burr models

Sample size	100	200	500	1000	2000
$CTE_{0.05}$=1.6970		$\xi = 2/3$ and $\rho = -0.5$			
$p = 0$	1.1854	1.3256	**1.6396**	**1.6922**	1.7518
$p = 0.5$	1.3663	**1.5344**	1.5969	1.7028	**1.7006**
$p = 0.75$	**1.4407**	1.5263	1.6054	1.6426	1.6860
$p = 1$	1.3913	1.4127	1.4935	1.5361	1.5807
$p = 1.25$	1.3081	1.3626	1.4551	1.4974	1.5520
$CTE_{0.05}$=2.5413		$\xi = 2/3$ and $\rho = -1$			
$p = 0$	**2.5217**	2.5837	2.6839	2.6327	2.6108
$p = 0.5$	2.4460	**2.5396**	2.5186	2.6389	2.5907
$p = 0.75$	2.4112	2.4866	**2.5547**	**2.5992**	**2.5627**
$p = 1$	2.2414	2.3114	2.3520	2.3938	2.4256
$p = 1.25$	2.1722	2.2549	2.3108	2.3512	2.3982
$CTE_{0.05}$=1.0520		$\xi = 0.5$ and $\rho = -0.5$			
$p = 0$	**1.0474**	1.0665	1.0669	1.0671	1.0600
$p = 0.5$	1.0305	**1.0529**	1.0582	1.0678	**1.0560**
$p = 0.75$	1.0325	1.0147	**1.0478**	**1.0525**	1.0566
$p = 1$	0.9961	1.0040	1.0178	1.0365	1.0395
$p = 1.25$	0.9725	0.9868	1.0069	1.0322	1.0342
$CTE_{0.05}$=1.6455		$\xi = 0.5$ and $\rho = -1$			
$p = 0$	1.6821	1.6655	1.6604	1.6564	1.6548
$p = 0.5$	1.6666	**1.6514**	**1.6464**	**1.6549**	1.6578
$p = 0.75$	**1.6271**	1.6329	1.6431	1.6659	1.6572
$p = 1$	1.5923	1.6234	1.6282	1.6354	**1.6400**
$p = 1.25$	1.5645	1.5988	1.6116	1.6202	1.6278

Table 2 Simulated medians at the simulated optimal level for GP models

Sample size	100	200	500	1000	2000
$CTE_{0.05}$=4.2092		$\xi = 3/4$			
$p = 0$	2.9379	3.2987	**4.2211**	4.0590	4.0843
$p = 0.5$	3.1402	3.5901	3.8260	4.1034	**4.1655**
$p = 0.75$	**3.5140**	**3.6111**	4.0399	**4.2946**	4.1462
$p = 1$	3.0173	3.1257	3.2281	3.1838	3.2110
$p = 1.25$	2.9244	3.0378	3.1912	3.1723	3.2075
$CTE_{0.05}$=3.1565		$\xi = 2/3$			
$p = 0$	2.6244	2.7691	2.7633	3.1007	**3.1533**
$p = 0.5$	**2.7360**	2.7407	3.0251	3.1049	3.1452
$p = 0.75$	2.7134	**2.9130**	**3.1011**	**3.1534**	3.1911
$p = 1$	2.6038	2.6635	2.6766	2.6701	2.6729
$p = 1.25$	2.5169	2.6063	2.6604	2.6640	2.6712
$CTE_{0.05}$=2.1039		$\xi = 0.5$			
$p = 0$	**2.0734**	2.0202	1.9902	2.0858	2.0633
$p = 0.5$	2.0202	1.9927	2.1171	2.1113	2.1243
$p = 0.75$	1.9962	**2.0255**	**2.1024**	**2.1037**	**2.0996**
$p = 1$	1.9559	1.9590	1.9671	1.9695	1.9796
$p = 1.25$	1.8971	1.9481	1.9620	1.9666	1.9786

Table 3 Simulated median square error at the simulated optimal level for Burr models

Sample size	100	200	500	1000	2000
$\xi = 2/3$ and $\rho = -0.5$					
$p = 0$	0.4749	0.2603	0.1427	0.0768	0.0464
$p = 0.5$	0.3348	0.2048	0.1131	0.0630	0.0409
$p = 0.75$	0.2711	0.1743	0.0953	0.0546	0.0349
$p = 1$	0.2364	0.1540	0.0873	0.0575	0.0385
$p = 1.25$	**0.2124**	**0.1450**	**0.0803**	**0.0518**	**0.0319**
$\xi = 2/3$ and $\rho = -1$					
$p = 0$	0.3712	0.2311	0.0937	0.0493	0.0283
$p = 0.5$	0.2876	0.1962	0.0801	**0.0408**	0.0254
$p = 0.75$	0.2410	0.1720	**0.0688**	0.0414	**0.0246**
$p = 1$	0.2407	0.1686	0.0876	0.0559	0.0354
$p = 1.25$	**0.2181**	**0.1452**	0.0842	0.0507	0.0339
$\xi = 0.5$ and $\rho = -0.5$					
$p = 0$	0.0339	0.0173	0.0078	0.0039	0.0019
$p = 0.5$	0.0260	0.0155	0.0071	0.0036	**0.0018**
$p = 0.75$	0.0232	0.0140	0.0064	0.0033	0.0019
$p = 1$	0.0202	0.0134	0.0065	0.0032	0.0019
$p = 1.25$	**0.0190**	**0.0118**	**0.0062**	**0.0030**	**0.0018**
$\xi = 0.5$ and $\rho = -1$					
$p = 0$	0.0324	0.0170	0.0063	0.0036	0.0017
$p = 0.5$	0.0288	0.0167	**0.0060**	**0.0034**	**0.0016**
$p = 0.75$	0.0262	0.0158	**0.0060**	0.0035	0.0018
$p = 1$	0.0266	0.0148	0.0064	0.0038	0.0021
$p = 1.25$	**0.0248**	**0.0134**	0.0063	0.0035	0.0020

Table 4 Simulated median square error at the simulated optimal level for GP models

Sample size	100	200	500	1000	2000
$\xi = 3/4$					
$p = 0$	2,.6081	1.4738	0.7806	0.4515	0.2073
$p = 0.5$	2.0059	1.0917	0.,5261	0.2986	0.1365
$p = 0.75$	1.6313	**0.9251**	**0.3864**	**0.2408**	**0.1072**
$p = 1$	**1.5429**	1.1953	0.9624	1.0513	0.9963
$p = 1.25$	1.7286	1.3862	1.0362	1.0751	1.0034
$\xi = 2/3$					
$p = 0$	0.7267	0.4528	0.1909	0.1042	0.0521
$p = 0.5$	0.5804	0.3233	0.1482	0.0773	0.0380
$p = 0.75$	0.5129	**0.2871**	**0.1202**	**0.0591**	**0.0305**
$p = 1$	**0.4567**	0.2987	0.2337	0.2367	0.2339
$p = 1.25$	0.4836	0.3327	0.2470	0.2426	0.2356
$\xi = 0.5$					
$p = 0$	0.1165	0.0428	0.0229	0.0109	0.0060
$p = 0.5$	0.0959	0.0403	0.0204	0.0099	0.0053
$p = 0.75$	0.0838	**0.0398**	**0.0193**	**0.0089**	**0.0048**
$p = 1$	0.0808	0.0411	0.0252	0.0190	0.0157
$p = 1.25$	**0.0804**	0.0438	0.0267	0.0196	0.0158

parent($\xi = 2/3, \rho = -0.5$) the CTE-estimator in (8) has the smallest absolute median-bias for sample sizes n=500, 1000. The smallest median square error is always achieved by $\mathbb{C}_{k,n,p,q}(\underline{\mathbf{X}}_n)$ in (9), with $p \neq 0$. For the Burr models under study, when $\rho = -0.5$, the best choice for p is the value 1.25. When $\rho = -1$ and $n \geq 500$, the values $p = 0.5, 0.75$ are the best ones. For the GP parents presented and for samples sizes $n \geq 200$ the choice $p = 0.75$ is the one providing the smallest median square errors.

4 Concluding Remarks

Rather than using the Hill EVI-estimator, H(k), for the semi-parametric estimation of the CTE, it seems sensible to use the power-mean-of-order-p EVI-estimator. Indeed, for $q = 0.05$ and for the models considered in this paper, it is always possible to find a positive value of p that allows a reduction in the estimator's median squared error and a reduction in the median-bias. Non-regular frameworks, out of the scope of this article, deserve further attention.

Acknowledgments This project was partially financed by FCT – Fundação para a Ciência e a Tecnologia, through national funds, within the projects UIDB/0006/2020, UIDP/0006/2020 (CEA/UL) and UIDB/04674/2020, UIDP/04674/2020 (CIMA), and by HiTEc COST Action CA21163

References

1. Landsman, Z., Valdez, E.: Tail conditional expectations for elliptical distributions. N. Am. Actuarial J. **7**, 55–71 (2003). https://doi.org/10.1080/10920277.2003.10596118
2. Artzner, P., Delbaen, F., Eber, J.-M., Heath, D.: Coherent measures of risk. Math. Finance **9**, 203–228 (1999). https://doi.org/10.1111/1467-9965.00068
3. Gnedenko, B.V.: Sur la distribution limite du terme maximum d'une série aléatoire. Ann. Math. **44**, 423–453 (1943). https://doi.org/doi:10.2307/1968974
4. Hill, B.M.: A simple general approach to inference about the tail of a distribution. Ann. Stat. **3**, 1163–1174 (1975). https://doi.org/10.1214/aos/1176343247
5. Caeiro, F., Gomes, M.I., Beirlant, J., de Wet, T.: Mean-of-order p reduced-bias extreme value index estimation under a third-order framework. Extremes **19**(4), 561–589 (2016). https://doi.org/10.1007/s10687-016-0261-5
6. Penalva, H., Caeiro, F., Gomes, M.I., Neves, M.M.: An Efficient Naive Generalisation of the Hill Estimator: Discrepancy between Asymptotic and Finite Sample Behaviour. Notas e Comunicações CEAUL 02/2016 (2016). http://ceaul.org/wp-content/uploads/2018/10/NotaseCom-2.pdf
7. Penalva, H., Gomes, M.I., Caeiro, F., Neves, M.M.: A couple of non reduced bias generalized means in extreme value theory: an asymptotic comparison. Revstat-Stat. J. **18**(3), 281–298 (2020). https://www.ine.pt/revstat/pdf/REVSTAT_v18-n3-3.pdf. https://doi.org/10.57805/revstat.v18i3.301
8. Penalva, H., Gomes, M.I., Caeiro, F., Neves, M.M.: Lehmer's mean-of-order-p extreme value index estimation: a simulation study and applications. J. Appl. Stat. **47**(13–15) (Advances in Computational Data Analysis), 2825–2845 (2020). https://doi.org/10.1080/02664763.2019.1694871

9. Paulauskas, V., Vaičiulis, M.: A class of new tail index estimators. Ann. Inst. Stat. Math. **69**(2), 461–487 (2017). https://doi.org/10.1007/s10463-015-0548-3
10. Cabral, I., Caeiro, F., Gomes, M.I.: On the comparison of several classical estimators with a minimum variance reduced bias estimator of the extreme value index. Commun. Stat. Theory Methods **51**(1), 179–196 (2022). https://doi.org/10.1080/03610926.2020.1746970
11. Brilhante, M.F., Gomes, M.I., Pestana, D.: A simple generalisation of the Hill estimator. Comput. Stat. Data Anal. **57**(1), 518–535 (2013). https://doi.org/10.1016/j.csda.2012.07.019
12. Paulauskas, V., Vaičiulis, M.: On the improvement of Hill and some other estimators. Lith. Math. J. **53**, 336–355 (2013). https://doi.org/10.1007/s10986-013-9212-x
13. Beran, J., Schell, D., Stehlík, M.: The harmonic moment tail index estimator: asymptotic distribution and robustness. Ann. Inst. Stat. Math. **66**, 193–220 (2014). https://doi.org/10.1007/s10463-013-0412-2
14. Segers, J.: Residual estimators. J. Stat. Plann. Inference **98**(1–2), 15–27 (2001). https://doi.org/10.1016/s0378-3758(00)00321-9
15. Brazauskas, V., Jones, B., Puri, M., Zitikis, R.: Estimating conditional tail expectation with actuarial applications in view. J. Stat. Plann. Inference **138**, 3590–3604 (2008). https://doi.org/10.1016/j.jspi.2005.11.011
16. Necir, A., Rassoul, A., Zitikis, R.: Estimating the conditional tail expectation in the case of heavy-tailed losses. J. Probab. Stat. (2010). https://doi.org/10.1155/2010/596839
17. Deme, E.H., Girard, S., Guillou, A.: Reduced-bias estimator of the conditional tail expectation of heavy-tailed distributions. In: Hallin, M., Mason, D., Pfeifer, D., Steinebach, J. (eds.) Mathematical Statistics and Limit Theorems. Springer, Cham (2015). https://doi.org/10.1007/978-3-319-12442-1_7
18. Laidi, M., Rassoul, A., Old Rouis, H.: Improved estimator of the conditional tail expectation in the case of heavy-tailed losses. Stat. Optim. Inf. Comput. **8**(1), 98–109 (2020). https://doi.org/10.19139/soic-2310-5070-665
19. Weissman, I.: Estimation of parameters and larger quantiles based on the k largest observations. J. Am. Stat. Assoc. **73**, 812–815 (1978). https://doi.org/10.2307/2286285
20. Haan, L. de: Slow variation and characterization of domains of attraction. In: Tiago de Oliveira, J. (ed.) Statistical Extremes and Applications, pp. 31–48. D. Reidel, Dordrecht (1984). https://doi.org/10.1007/978-94-017-3069-3_4
21. Hall, P., Welsh, A.H.: Adaptive estimates of parameters of regular variation. Ann. Stat. **13**, 331–341 (1985). https://doi.org/10.1214/aos/1176346596
22. Gomes, M.I., Henriques-Rodrigues, L., Pestana, D.: Estimação de um índice de valores extremos positivo através de médias generalizadas e em ambiente de não-regularidade. In: Milheiro, P., et al. (eds.) Estatística: Desafios Transversais às Ciências com Dados—Atas do XXIV Congresso da Sociedade Portuguesa de Estatística, Edições SPE, pp. 213–226 (2021). https://www.spestatistica.pt/storage/app/uploads/public/609/28f/6d0/60928f6d08a0c016386627.pdf
23. Gomes, M.I., Henriques-Rodrigues, L., Pestana, D.: A generalized mean under a non-regular framework and extreme value index estimation. In: Zafeiris, K.N., Dimotikalis, Y., Skiadas, C.H., Karagrigoriou, A., Karagrigoriou-Vonta, C. (eds.) Data Analysis and Related Applications 2, Volume 10—Big Data, Artificial Intelligence and Data Analysis, iSTE Wiley, Part 3, Chapter 16, 237–250 (2022). https://www.iste.co.uk/book.php?id=1928
24. Gomes, M.I., Henriques-Rodrigues, L., Pestana, D.: Non-regular frameworks and the mean-of-order p extreme value index estimation. J. Stat. Theory Practice **16**, 37 (2022). https://doi.org/10.1007/s42519-022-00264-w
25. Nadarajah, S., Chan, S., Afuecheta, E. VaRES: Computes value at risk and expected shortfall for over 100 parametric distributions. R package version 1.0. (2013). https://cran.r-project.org/web/packages/VaRES/VaRES.pdf
26. Chan, S., Nadarajah, S., Afuecheta, E. An R package for value at risk and expected shortfall. Commun. Stat. Simul. Comput. **45**(9), 3416–3434, (2016). https://doi.org/10.1080/03610918.2014.944658
27. R Core Team (2021). R: A language and environment for statistical computing. R Foundation for Statistical Computing, Vienna. https://www.R-project.org/

Risk Analysis in Practice and Theory

Christos P. Kitsos

Abstract In this paper we discuss the Risk Analysis problem as it has been developed, offering a solution to crucial problems and offering food for thought for statistical generalisations. We try to explain why we need to keep the balance between Theory and Practice.

Keywords Risk Analysis · Hazard function · Generalised normal distribution

1 Introduction

At the early stage risk was involving to political or military games for a decision making with the minimum risk. The pioneering work of Quincy Wright [40] on the study of war was devoted to this line of thought. The Mathematics and Statistics involved, could be considered in our day as low-level, he applied eventually the differential equation theory with a successful application.

In principle Risk is defined as an exposure to the chance of injury or loss—it is a hazard or dangerous chance for an event under consideration. Therefore the probability of a damage, for the considered phenomenon (in Politics, Economy, Epidemiology, Food Science, Industry etc.) caused by external or internal factors has to be evaluated, especially the essential ones influence the Risk. That is why we refer eventually to Relative Risk (RR), as each factor influences the Risk in a different way. In principle the relative risk (RR) is the ratio of the probability of an outcome in an exposed group to the probability of an outcome in an unexposed group. That is why a value of RR = 1 means that the exposure does not affect the outcome and a "risk factor" is assigned when RR> 1, i.e. when the risk of the outcome is increased by the exposure.

C. P. Kitsos (✉)
Department of Informatics and Computer Engineering, University of West Attica, Egaleo, Greece
e-mail: xkitsos@uniwa.gr

© The Author(s), under exclusive license to Springer Nature Switzerland AG 2023

C. P. Kitsos et al. (eds.), *Statistical Modelling and Risk Analysis*,
Springer Proceedings in Mathematics & Statistics 430,
https://doi.org/10.1007/978-3-031-39864-3_9

This is clear in Epidemiological studies where in principle it is needed to identify and quantitatively assess the susceptibility of a partition of the population to specify risk factors, so we refer to RR. For a nice introduction to statistical terminology for RR see Everitt [13, Chapter 12].

Such an early attempt was by John Graunt (1620–1674), founder of Demography, trying to evaluate "bills of Mortality" as he explained in his work "Observations", while almost at the same time his friend Sir William Petty (1623–1687), economist and Philosopher, published the "Political Anatomy of Ireland". So there was an early attempt to evaluate Social and Political Risk.

Still there is a line of thought loyal to the idea that Risk Analysis is only related to political problems through the Decision Theory; William Playfair (1759–1823) was among the first working with empirical data in 1796 publishing "For the Use of the Enemies of England: A Real Statement of the Finances and Resources of Great Britain". Quincy Wright (1890–1970) in his excellent book "A study of War" offers a development of simple indexes, evaluating Risk successfully, as it has been pointed out for such an important problem as war.

The statistical work of Florence Nightingale (1820–1910) is essential as with her "Notes on Matters effecting the Health, Efficiency and Hospital Administration of the British Army" opened the problem of analysing Epidemiological Data, adopting the Statistical methods of that time.

It is really Armitage and Doll [2] who introduced the recent Statistical framework to the Cancer Problem. Latter Crump et al. [11] can be referred for their work on carcinogenic process, while [14] provided a global work for the Bioassays, and Megill [34] worked on Risk Analysis (RA) for Economical Data. It was emphasising that, at the early stages, the fundamental in RA was to isolate the involved variables. Still the Statistical background was not too high. But the adoption of the triangle distribution was essentially useful. The triangle distribution has been faced under a different statistical background recently, but still the triangle obtained from the mode, the minimum value its high and the maximum value of the data can be proved very useful, as a special case of trapezoidal distributions, see also Appendix 1. For a compact new presentation, while a more general framework was developed by Ngunen and McLachian [35]. The main characteristic of the triangular distribution is its simplicity and can be easily adopted in practice. There are excellent examples with no particular mathematical difficulty in Megill [34].

In Food Science the Risk Assessment problem is easier to be understood by those who are not familiar to RA. In the next section we discuss the existing Practical Background, which is not that easy to be developed, despite the characterisation as "practical". Most of the ideas presented are from the area of Food Science where RA is very clear under a chemical analysis orientation.

In Sect. 3 the existing theoretical insight is discussed briefly and therefore the Discussion in Sect. 4 is based on Sects. 2 and 3.

2 Practical Background

Risk factors can be increased during the food processing and food can be contaminated due to filtering and cleaning agents or during packaging and storage. Therefore, in principle, chemical hazards can be divided in two primary categories:

(i) Naturally occurring chemical hazards (mycotoxins, pyrrolizidine, alkaloids, polychlorinated biphenyls etc.)
(ii) Added chemical hazards (pesticides, antibiotics, hormones, heavy metals etc.)

The effect of each chemical as a Risk factor has been studied and we refer briefly to mycotoxins as dairy products belong to the most susceptible foodstuffs (one possible reason humidity, among others) to be contaminated by them and might result to, Kitsos and Tsaknis [28] among others.

1. Direct contamination
2. Indirect contamination

Example 1 (Indirect Contamination) Recall that due to decontamination bacteria become resistant and therefore interhospital various can appear. Moreover a number of countries have introduced or proposed regulations for the control and analysis of aflatoxins in food.

As far as milk is concerned, EU requires the maximum level of aflatoxin M_1, max M_1 say, max $M_1 = 0.5$ mg/kg. The maximum tolerated level for aflatoxin M_1, in dairy products, it is not the same all over the world and therefore it is regulated in some countries.

The problem of mixtures has been discussed from a statistical point of view, for the cancer problem, in Kitsos and Elder [23]. In practice the highly carcinogenic polychlorinated biophenyls (PCBs) are mixtures of chlorinated biphenyls with varying percentages of chlorine/weight. It has been noticed, Biuthgen et al. [3], that PCBs led to a worldwide contamination of the environment due to their physical/chemical properties. Moreover PCBs have been classified as probable human carcinogens, while no Tolerance Daily Intake (TDI), the main safety standards, have been established for them. Eventually the production of PCBs was banned in USA in 1979 and internationally in 2001.

Example 2 Dioxins occur as complex mixtures, Kitsos and Edler [23], and mixtures act through a common mechanism but vary in their toxic potency. As an example Tetrachlorodibenzo-p-dioxin (TCCD) has been classified as a human carcinogen, as there are epidemiological studies on exposure to 2,3,7,8- tetrachlorodizen-p-dioxin and cancer risk. It might not be responsible for producing substantial chronic disability in humans but there are experimental evidence for its carcinogenicity, McConnell et al. [33].

The TDI for dioxins is 1–4 pg TEQ/kg body-weight/day, which is exceeded in industrialised countries. Recall that Toxic Equivalence Quotient (TEQ) is the USA Environmental Protection Agency (EPA),with TEQ being the, threshold for safe dioxin exposure at Toxicity Equivalence of 0.7 picograms per kilogram of body weight per day.

The Lethal Dose is an index of the percentage P of the lethal toxicity LD_P of a given toxic substance or different type radiation. $LD_{0.5}$ is the amount of given material at once that causes the death of 50% in the group of animals (usually rats and mice) under investigation. Furthermore the median lethal dose $LD_{0.5}$ is widely used as a measure of the same effect in toxic studies. Not only the lethal dose but also the low percentiles need special consideration, see Kitsos [18], who suggested a sequential approach to face the problem.

Now if we assume that two components C_1 and C_2 are identical, except that C_1 is thinned by a factor $T < 1$, then we can replace the same dose as d_1 of C_1 by an appropriate dose of C_2, so that to have the same effect as dose d_1. In such a case the effect of a combination of doses d_1 and d_2, for the components under consideration, C_1 and C_2 are: $Td_1 + d_2$ of C_2, $d_1 + (1/T)d_2$ of C_1 respectively. The factor T is known as relative potency of C_1 to C_2 and $\lambda = 1/T$ is called as relative potency of C_2 to C_1. Such simple but practical rules are appreciated by experimenters. Another practical problem in RA appears with the study of the involved covariates. The role of covariates, in this context, is of great interest and has been discussed by Kitsos [17].

Therefore, in principle, to cover as many as possible sources of risk as possible, we can say that the target in human risk assessment is the estimation of the probability of an adverse effect to human being, and the identification of such a source.

3 Theoretical Inside

In Biostatistics and in particular in Risk Analysis for the Cancer problem, the evolution of the Statistical applications can be considered in the over 1000 references in Edler and Kitsos [12]. The development of methods and the application of particular probabilistic models, Kopp-Schneider et al. [30] and statistical analyses appear on extended development after 1980. Recently, Stochastic Carcinogenesis Models, Dose Response Models on Modeling Lung Cancer Screening are medical ideas with a strong statistical insight which have been adopted by the scientific community, Kitsos [21].

The variance-covariance matrix is related to Fisher's information matrix and it is the basis for evaluating optimal designs in chemical kinetics, Kitsos and Kolovos [24], while for a recent review of the Mathematical models, facing breast cancer see Mashekova et al. [32]. The Fisher's information measure appears either in a parametric form, or in an entropy type. The former plays an important role to a number of Statistical applications, such as the optimal experimental design, the calibration problem, the variance estimation of the parameters in Logit model in RA, Cox [9, 10], etc. The latter is fundamental to the Information Theory. Both have been extended by Kitsos an Tavoularis [26].

Indeed: With the use of an extra parameter, which influences the "shape" of the distribution, the generalised Normal distribution was introduced. This is useful in cases where "fat tails" exist, i.e. the Normal distribution devotes 0.05 probability in details but there are cases where the distribution provides in "tails" more than 0.05 probability. Such cases are covered under the generalised Normal distribution.

The γ-ordered Normal Distribution emerged from the Logarithm Sobolev Inequality and it is a generalisation of the multivariable Normal distribution, with an extra parameter γ in the following, see also Appendix 2. It can be useful to RA to adopt the general cumulative hazard function, see (2), (3) below. Therefore a strong mathematical background exists, which is certainly difficult to be followed by toxicologists, medical doctors, etc who mainly work on RA. Still it has not been developed an appropriate software for it.

The Normal distribution has been extended by Kitsos and Tavoularis [26], with a rather complicated form, quite the opposite of the easy to handle the triangular distribution, see Appendix 1.

Let, as usual, $\Gamma(a)$ be the gamma function and $\Gamma(x, a)$ the upper incomplete gamma function. Then the cumulative distribution function (cdf) of the $GN(\mu, \sigma^2; \gamma)$, say,

$$\Phi_G(x) = 1 - \frac{\Gamma(\gamma_0, \gamma_0 z^{\frac{1}{\gamma_0}})}{2\Gamma(\gamma_0)}, \quad \gamma_0 = \frac{\gamma - 1}{\gamma}, \quad \gamma \in \mathbb{R} - [0, 1], \quad z = \frac{x - \mu}{\sigma} \tag{1}$$

with μ the position parameter, σ the scale parameter and γ an extra, shape parameter. In this line of thought Kitsos and Toulias [25] as well as Toulias and Kitsos [37] worked on the Generalised Normal Distribution $GN(\mu, \sigma^2; \gamma)$ with $\gamma \in \mathbf{R} - [0, 1]$ being an extra shape parameter. This extra parameter γ makes the difference: when $\gamma = 2$ the usual Normal is obtained, with $\gamma > 0$ it is still normal with "heavy tails".

Under this foundation the cumulative hazard function, $H(\cdot)$ say, of a random variable $X \sim GN(\mu, \sigma^2; \gamma)$ can be proved equal to

$$H(x) = -\log A(\gamma_0, z), \quad x > \mu \tag{2}$$

while

$$H(x) = -\log(1 - A(\gamma_0, |z|)), \quad x \leq \mu \tag{3}$$

with

$$z = \frac{x - \mu}{\sigma}, \quad A(\gamma_0, z) = \frac{\Gamma(\gamma_0, \gamma_0 z^{\frac{1}{\gamma_0}})}{2\Gamma(\gamma_0)}.$$

with $\Gamma(a)$ being the gamma function and $\Gamma(x, a)$ the upper incomplete gamma function.

Example 3 As $\gamma \to \pm\infty$ the Generalised Normal Distribution tends to Laplace, $L(\mu, \sigma)$. Then it can be proved that:

$$H(x) = \log(2 + \frac{x - \mu}{\sigma})\,, \quad x > \mu \tag{4}$$

while

$$H(x) = \log(1 - \frac{1}{2} e^{\frac{x-\mu}{\sigma}})\,, \quad x \leq \mu\,. \tag{5}$$

See Toulias and Kitsos [37] for more examples.

Let X be the rv denoting the time of death. Recall that the future lifetime of a given time x_0 is the remaining time $X - x_0$ until death. Therefore the expected value, $E(X)$, of the future life time can be evaluated . In principle it has to be a function of the involved survival function, Breslow and Day [5]. This idea can be extended with the γ-order Generalised Normal. Moreover for the future lifetime rv X_0 at point x_0, $X \sim GN(\mu, \sigma^2; \gamma)$ the density function (df), the cdf can be evaluated and the corresponding expected future lifetime is

$$E(X) = \frac{2(\mu - x_0)}{A(\gamma_0, z_0)}\,, \quad z_0 = \frac{x_0 - \mu}{\sigma}\,. \tag{6}$$

The above mentioned results, among others, provide evidence to discuss, that the theoretical inside is moving faster than the applications are needed such results. These comments need special consideration and further analysis. We try in Sect. 4.

4 Discussion

To emphasise how difficult the evaluation of Risk might be, we recall the Simpson's paradox, Blyth [4], when three events A, B, C are considered. Then if we assume

$$\begin{aligned} P(A|BC) &\geq P(A|\bar{B}C)\,, \\ P(A|B\bar{C}) &\geq P(A|\bar{B}\bar{C})\,, \end{aligned} \tag{7}$$

we might have

$$P(A|B) \leq P(A|\bar{B})\,. \tag{8}$$

Therefore there is a prior, a scepticism of how "sure" a procedure might be. In Epidemiological studies it is needed to identify and quantitatively assess the susceptibility of a portion of the population to specific Risk factors. It is assumed that they have been exposed to the *same possible hazardous factors*. The difference that at the early stage of the study, is only on a particular factor which acts as a susceptibility factor. In such cases Statistics provides the evaluation of the RR. That is why J. Grant was mentioned in Sect. 1.

Concerning the 2 × 2 setup, for correlated binary response, the backbone of medical doctors research, a very practical line of thought, with a theoretical background was faced by Mandal et al. [31] and is exactly the spirit we would like to encourage, following Cox believes, Kitsos [22] of how Statistics can support other Sciences, especially medicine. They provide the appropriate proportions and their variances in a 2 × 2 setup, so that 95% confidence interval can be constructed. The Binary Response problem has been early discussed by Cox [9] while for a theoretical approach for Ca problems see Kitsos [20].

The area of interest of RA is wide; it covers a number of fields, with completely different backgrounds sometimes, such as Politics and food Science. But Food Science is related to Cancer problems as we briefly discussed.

Excellent Economical studies with "elementary" statistical work are covered by Megill [34] who provides useful results as Wright [40] did earlier. Therefore we oscillate between Practice and Theory. We have theoretical results, waiting to be applied as in the 60s we had Cancer problems waiting for statistical considerations.

The cancer problem was eventually the problem under consideration and Sir David Cox provided a number of examples working on this, Cox [8–10], and offers ideas of how we can proceed on medical data analysis, Kitsos [22], trying to keep it simple. In contrast Tan [36], offers a completely theoretical approach, understanding perhaps from mathematicians, Kopp-Schneider [29] reviews the theoretical stochastic models and in lesser extent Kitsos [19, 21], Kitsos and Toulias [25] the appropriate modeling, which are difficult to be followed by medical doctors and not only.

A compromise between theory and practice has been attended in Edler and Kitsos [12],where different approaches facing cancer are discussed, while Cogliano et al. [7] discuss more toxicological oriented cases. The logit method took some time to be appreciated, but provides a nice tool for estimating Relative Risk, Kitsos [20], among others. The role of covariates in such studies, and not only for cancer it is of great interest and we believe is needed to be investigated, Kitsos [17]. In this paper we provided food for thought for a comparison of an easy to understand work with the triangular distribution and the rather complicated Generalised Normal, see Appendix 2. It is not only a matter of choice. It depends heavily on the structure of data—we would say graph your data and then proceed your analysis.

The logit methods can be applied on different applied areas. Certain qualities have been adopted for different areas from international organisations, see IARC [16], WHO [39], US EPA [38] among others. As it is mentioned in Sect. 2, as far as Food Risk Assessment concerns, Fisher et al. [15], Kitsos and Tsaknis [28], Binthgen et al. [3] among others, there are more chemical results and guidance for the involved risk, while Amaral-Mendes and Pluyger [1] offer an extensive list of Biomarkers for Cancer Risk Assessment in humans.

In Cancer problems, and not only, the hazard function identification is crucial and only Statistical Analysis can be adopted, Armitage and Doll [2], Crump et al.

[11], Cogliano et al. [7], Kitsos [17, 18]. The extended work, based on generalised Normal distribution, mentioned in Sect. 2 in a global form, generalising the hazard function, needs certainly not only an appropriate software cover but also to bridge the differences between statistical line of thought and applications.

Meanwhile recent methods can be applied to face cancer, Carayanni and Kitsos [6], where the existent software offers a great support. More geometrical knowledge is needed, or even fractals, to describe a tumour. But the communication with Medicine might be difficult.

We need to keep the balance of how "Statistics in Action" has to behave offering solutions to crucial problems of Risk Analysis, see Mandal et al. [31], while the theoretical work of Tan [36] adds a strong background but not useful to practical problems. Since the time that Cox [10] provided a general solution for hazard functions, there is an extensive development of Statistical Theory for Risk problems.

It might be eventually helpful to offer results, but now we believe it is also very crucial to offer solutions, to the corresponding fields, working in Risk Analysis. That is the practical background is needed, we believe, to be widely known, as it is easier to be absorbed from practician and the theoretical framework is needed to be supported from the appropriate software so that to bridge the gap with practical applications.

Acknowledgments I would like to thank the referees for their comments which improved the final version of this paper. The discussions with Associate Professor S. Fatouros are very much appreciated.

Appendix 1

Let $X_1, X_2, ..., X_n$ be a set of n independent, identically distributed, random variables with

$$m = \min\{X_i\},\ M_0 = mode\{X_i\},\ M = \max\{X_i\},\ i = 1, 2, ...n$$

with these three points m, M_0, M we can define a plane triangle with height $h = \frac{2}{M-m}$ and the points m, M_0, M on the basis of the triangle. Notice that for a continuous random variable the mode is not the value of X most likely to occur, as it is the case for discrete random variables. It is worth it to notice that the mode of a continuous random variable corresponds to that x value/values at which the probability density function (pdf) $f(\cdot)$ reaches a local maximum, or a peak, i.e. is the solution of the equation $\frac{df(x)}{dx} = 0$ See Megill [34], for the definitions and a Euclidean Geometry development.

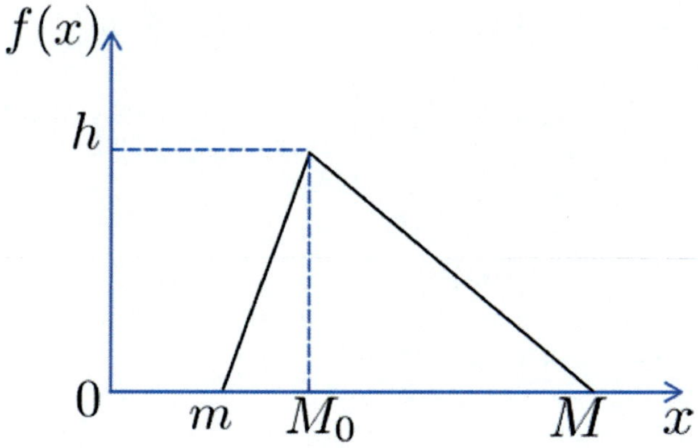

Fig. 1 The triangle distribution

The probability density function of the triangle distribution is defined as

$$f_t(x) = \begin{cases} h\frac{1}{d_1}(x - m) & x \in [m, M_0) \\ \\ h\frac{1}{d_2}(M - x) & x \in (M_0, M] \end{cases} \tag{9}$$

with $d_1 = M_0 - m$, $d_2 = M - M_0$.

If we let $v = (m, M_0, M)$ and $u = (M, m, M_0)$, $\mathbf{1} = (1, 1, 1)$, then

$$\begin{aligned} E(X) &= \tfrac{m+M_0+M}{3} = \tfrac{1}{3}\mathbf{1}v^T \\ V(X) &= \tfrac{1}{18}[vv^T - vu^T] = \tfrac{1}{18}[\|v\|^2 - < v, u >] \end{aligned} \tag{10}$$

with $\| \cdot \|$ the Euclidean norm and $< \cdot, \cdot >$ the inner product of two vectors (Fig. 1, see also Megiil [34]).

It is helpful that in triangle distribution the mode lies within the range $R \simeq 6\sigma$, σ being the standard deviation.

See also Nguyen and McLachlan [35] for a more general analysis for the triangle distribution.

Appendix 2

Let p be the number of parameters involved in the multivariate normal distribution. The induced from the Logarithm Sobolev Inequality (LSI), γ-ordered Normal distribution $GN^p(\gamma; \mu, \Sigma)$ behaves as a generalized multivariate normal distribution, with an extra parameter, with $\gamma \in \mathbb{R}$ and it is assumed that $\gamma_1 = \frac{\gamma}{\gamma-1} > 0$.

The density function of the γ-ordered Normal is defined as

$$f(x) = C(p, \gamma)|\det \Sigma|^{-1/2} \exp\left\{-\frac{\gamma - 1}{\gamma} Q(x)^{\frac{\gamma}{2(\gamma-1)}}\right\}, \ x \in \mathbf{R}^p \tag{11}$$

where

$$Q(x) = (x - \mu)\Sigma^{-1}(x - \mu)^T$$

and the normalizing factor equals to $C(p, \gamma)$ equals to

$$C(p, \gamma) = \pi^{-p/2} \frac{\Gamma(\frac{p}{2} + 1)}{\Gamma(p\frac{\gamma-1}{\gamma} + 1)} \left(\frac{\gamma - 1}{\gamma}\right)^{p\frac{\gamma-1}{\gamma}} . \tag{12}$$

Notice that, from the definition in (2) the second-ordered Normal is the known normal distribution, i.e. $GN^p(2; \mu, \Sigma) = N(\mu, \Sigma)$. In the spherically contoured case, i.e. when $\Sigma = \sigma^2 I_p$, the density f_γ is reduced to the form

$$f(x) = \frac{\Gamma\left(\frac{p}{2} + 1\right)\left(\frac{\gamma-1}{\gamma}\right)^{p\frac{\gamma-1}{\gamma}}}{\Gamma\left(p\frac{\gamma-1}{\gamma} + 1\right)\pi^{p/2}\sigma^p} \exp\left[-\frac{\gamma - 1}{\gamma}\left(\frac{|x - \mu|}{\sigma}\right)^{\frac{\gamma}{\gamma-1}}\right], \ x \in \mathbb{R}^p . \tag{13}$$

For a random variable X following $GN^p(\gamma, \mu, \sigma^2 I_p)$ we can evaluate its mode, Mode(X), which is achieved due to the symmetry as in classical Normal distribution, for $x = \mu$, i.e.

$$\text{Mode}(X) = \frac{\left(\frac{\gamma-1}{\gamma}\right)^{p\frac{\gamma-1}{\gamma}}}{\pi^{p/2}\sigma^p} \frac{\Gamma\left(\frac{p}{2} + 1\right)}{\Gamma\left(p\frac{\gamma-1}{\gamma} + 1\right)} . \tag{14}$$

which can be easily verified for the classical normal with $\gamma = 2$ and $p = 1$ (single variable), recall also the symmetry, see Kitsos and Toulias [27] for details.

Theorem 1 (Kitsos and Toulias [27]) *The spherically contoured γ-order Generalised Normal distribution, coincides with the p-variate normal distribution when $\gamma = 2$, with the p -variate uniform distribution when $\gamma = 1$, and with the p-variate Laplace distribution when $\gamma = \pm\infty$.*

References

1. Amaral-Mendes, J.J., Pluygers, E.: Use of biochemical and molecular biomarkers for cancer risk assessment in humans. In: Perspectives on Biologically Based Cancer Risk Assessment, pp. 81–182. Springer, Boston (1999)
2. Armitage, P., Doll, R.: The age distribution of cancer and a multi-stage theory of carcinogenesis. Br. J. Cancer, **8**, 1–12 (1954)
3. Blüthgen, A., Burt, R., Heeschen, W.H.: Heavy metals and other trace elements. Monograph on Residues and Contaminants in Milk and Milk Products. IDF Special Issue 9071, 65–73 (1997)
4. Blyth, C.R.: On Simpson's paradox and the sure-thing principle. J. Am. Stat. Assoc. **67**(338), 364–366 (1972)
5. Breslow, N.E., Day, N.E.: Statistical methods in cancer research. Volume I-The analysis of case-control studies. IARC Sci. Publ. (32), 5–338 (1980)
6. Carayanni, V., Kitsos, C.: Model oriented statistical analysis for cancer problems. In: Pilz, J., Oliveira, T., Moder, K., Kitsos, C. (eds.) Mindful Topics on Risk Analysis and Design of Experiments, pp. 37–53. Springel, (2022)
7. Cogliano, V.J., Luebeck, E.G., Zapponi, G.A. (eds.): Perspectives on Biologically Based Cancer Risk Assessment (23). Springer Science & Business Media, Berlin (2012)
8. Cox, D.R.: Tests of separate families of hypotheses. In: Proceedings of the Fourth Berkeley Symposium on Mathematical Statistics and Probability 1(0), 96 (1961)
9. Cox, D.R.: Analysis of Binary Data. Chapman and Hall, London (1970)
10. Cox, D.R.: Regression models and life-tables. J. R. Stat. Soc. Ser. B Methodol. **34**(2), 187–202 (1972)
11. Crump, K.S., Hoel, D.G., Langley, C.H., Peto, R.: Fundamental carcinogenic processes and their implications for low dose risk assessment. Cancer Res. **36**(9 Part 1), 2973–2979 (1976)
12. Edler, L., Kitsos, C. (eds.) Recent Advances in Quantitative Methods in Cancer and Human Health Risk Assessment. Wiley (2005)
13. Everitt, B.S.: Statistical Methods for Medical Investigations, pp. 105–115. Edward Arnold, London (1994)
14. Finney, D.J.: Probit Analysis. Cambridge University Press, Cambridge (1971)
15. Fisher, W.J., Trisher, A.M., Schiter, B., Standler, R.J.: Contaminations of milk and dairy products. (A) Contaminants resulting from agricultural and dairy practices. In: Roginski, H., Fuquary, J.W., Fox, P.F. (eds.) Encyclopedia of Dairy Sciences. Academic Press, Oxford (2003)
16. IARC: Polychlorinated dibenzo-paradioxins and polychlorinated dibenzofurans. In: Monographs of the Evaluation of the Carcinogenic Risk of Chemicals to Humans (69). IARC, Lyon (1977)
17. Kitsos, C.P.: The role of covariates in experimental carcinogenesis. Biom. Lett. **35**(2), 95–106 (1998)
18. Kitsos, C.: Optimal designs for estimating the percentiles of the risk in multistage models of carcinogenesis. Biom. J.: J. Math. Methods Biosci. **41**(1), 33–43 (1999)
19. Kitsos, C.P.: The cancer risk assessment as an optimal experimental design. Folia Histochem. Cytobiol. Suppl. **39**(1), 16 (2001)
20. Kitsos, C.P.: On the logit methods for Ca problems. In: Vonta, F. (ed.) Statistical Methods for Biomedical and Technical Systems, pp. 335–340. Limassol, Cyprus (2006)
21. Kitsos, C.P.: Cancer Bioassays: A Statistical Approach. LAMBERT Academic Publisher, Saarbrucken, **326**, 110 (2012)
22. Kitsos, C.P.: Sir David Cox: a wise and noble Statistician (1924–2022). EMS Mag. **124**, 27–32 (2022)
23. Kitsos, C.P., Edler, L.: Cancer risk assessment for mixtures. In: Edler, L., Kitsos, C. (eds.) Recent Advances in Quantitative Methods in Cancer and Human Health Risk Assessment, pp. 283–298. Wiley, (2005)
24. Kitsos, C.P., Kolovos, K.G.: A compilation of the D-optimal designs in chemical kinetics. Chem. Eng. Commun. **200**(2), 185–204 (2013)

25. Kitsos, C.P., Toulias, T.L.: Generalized information criteria for the best logit model. In: Oliveira, T., Kitsos, C., Rigas, A., Gulati, S. (eds.) Theory and Practice of Risk Assessment, pp. 3–20. Springer, Cham (2015)
26. Kitsos, C.P., Tavoularis, N.K.: Logarithmic Sobolev inequalities for information measures. IEEE Trans. Inf. Theory **55**(6), 2554–2561 (2009)
27. Kitsos, C.P., Toulias, T.L.: On the family of the γ-ordered normal distributions. Far East J. Theor. Stat. **35**(2), 95–114 (2011)
28. Kitsos, C.P., Tsaknis, I.: Risk analysis on dairy products. In: Proceedings of the International Conference on Statistical Management of Risk Assessment, Lisbon 30–31 Aug (2007)
29. Kopp-Schneider, A.: Carcinogenesis models for risk assessment. Stat. Methods Med. Res. **6**(4), 317–340 (1997)
30. Kopp-Schneider, A., Burkholder, I., Groos, J.: Stochastic carcinogenesis models. In: Edler, L., Kitsos, C. (eds.) Recent Advances in Quantitative Methods in Cancer and Human Health Risk Assessment, pp. 125–135. Wiley, (2005)
31. Mandal, S., Biswas, A., Trandafir, P.C., Chowdhury, M.Z.I.: Optimal target allocation proportion for correlated binary responses in a 2×2 setup. Stat. Probab. Lett. **83**(9), 1991–1997 (2013)
32. Mashekova, A., Zhao, Y., Ng, E.Y., Zarikas, V., Fok, S.C., Mukhmetov, O.: Early detection of the breast cancer using infrared technology–a comprehensive review. Therm. Sci. Eng. Prog. **27**, 101142 (2022)
33. McConnell, E.E., Moore, J.A., Haseman, J., Harris, M.W.: The comparative toxicity of chlorinated dibenzo-p-dioxins in mice and guinea pigs. Toxicol. Appl. Pharmacol. **44**(2), 335–356 (1978)
34. Megill, R.E.: An Introduction to Risk Analysis. Penn. Well. Pub. Co., Tulsa (1984)
35. Nguyen, H.D., McLachlan, G.J.: Maximum likelihood estimation of triangular and polygonal distributions. Comput. Stat. Data Anal. **102**, 23–36 (2016)
36. Tan, W.Y.: Stochastic Models for Carcinogenesis, vol. 116. CRC Press, Boca Raton (1991)
37. Toulias, T.L., Kitsos, C.P.: Hazard rate and future lifetime for the generalized normal distribution. In: Oliveira, T., Kitsos, C., Oliveira, A., Grilo, L. (eds.) Recent Studies on Risk Analysis and Statistical Modeling, pp. 165–180. Springer, Cham (2018)
38. US EPA: Help Manual for Benchmark Pose Software (2000)
39. World Health Organization: Guidelines for air quality (No. WHO/SDE/OEH/00.02). World Health Organization (2000)
40. Wright, Q.: A Study of War. University of Chicago Press, Chicago (1964)

On Some Consequences of COVID-19 in EUR/USD Exchange Rates and Economy

Zachary R. Kuenstler, Brennan C. Merley, Milan Stehlik, Jerzy Filus, Lidia Filus, Claudia Navarro-Villarroel, Jean Paul Maidana, and Felix Fuders

Abstract Here we analyze several economical variables which have been affected by the period of COVID-19. In particular, EUR/US exchange rates are addressed. Oil prices have been very volatile and many other economical variables have changed their behavior.

We show by application of statistical tests for normality, including QQ-plots and Shapiro-Wilk that exchange rates of EUR to US from 10/13/2019 to 4/9/2020 are substantially deviating from normality and outliers are present. It is clear that changes to the economical variables have been of interest.

Keywords Exchange rates · COVID-19 · ARIMA · Modelling

Z. R. Kuenstler · B. C. Merley
University of Iowa, Iowa City, IA, USA

M. Stehlik (✉) · C. Navarro-Villarroel
Instituto de Estadística, Universidad de Valparaíso, Valparaíso, Chile
e-mail: milan.stehlik@uv.cl

J. Filus
Department of Mathematics and Computer Science, Oakton Community College, Des Plaines, IL, USA

L. Filus
Department of Mathematics, Northeastern Illinois University, Chicago, IL, USA

J. P. Maidana
Facultad de Ingeniería, Universidad Andrés Bello, Viña del Mar, Chile

Instituto de Estadística, Universidad de Valparaíso, Valparaíso, Chile

F. Fuders
Economics Institute, Universidad Austral de Chile, Valdivia, Chile

© The Author(s), under exclusive license to Springer Nature Switzerland AG 2023
C. P. Kitsos et al. (eds.), *Statistical Modelling and Risk Analysis*,
Springer Proceedings in Mathematics & Statistics 430,
https://doi.org/10.1007/978-3-031-39864-3_10

1 Exchange Rates 10/13/2019–4/9/2020

An exchange rate is the value of a currency in a given country compared with another country. They will rise or fall based on a country's supply and demand of imports/exports. It is one of the most useful tools to measure an economic growth or decline in the country.

We use exchange rates and other economic indicators to measure and predict the effect of COVID-19 on the United States economy. Abrupt changes of economic variables have been previously studied in Stehlik et al. [1], and also in [2] and [3].

Taking a quicklook on the original data, it is obvious that something significant happened between January and March, as it is show in Fig. 1. There are significant peaks and valleys that are shown in succession that must be due to an extraneous force in the economy: COVID-19. At the beginning of February was when the World Health Organization reconvened the Emergency Committee and assessed a very high global risk level.

If we examine the quantile-quantile plot of the exchange rates of EUR to USD from 2019/10/14 to 2020/4/9 in Fig. 2, we can see there is a strange object in the lower left portion that is very unique. Along with it, the Shapiro-Wilk test of normality gives a value of 0.94648 and a p-value of 1.236e−05. And the Jarque-Bera with 2 degrees of freedom gives a χ^2 value of 12.727, and a p-value of 0.001723; all showing strong non-normality. After identifying specific observations in the QQ-plot of Fig. 2, we can see that nearly every non-normal plot comes after

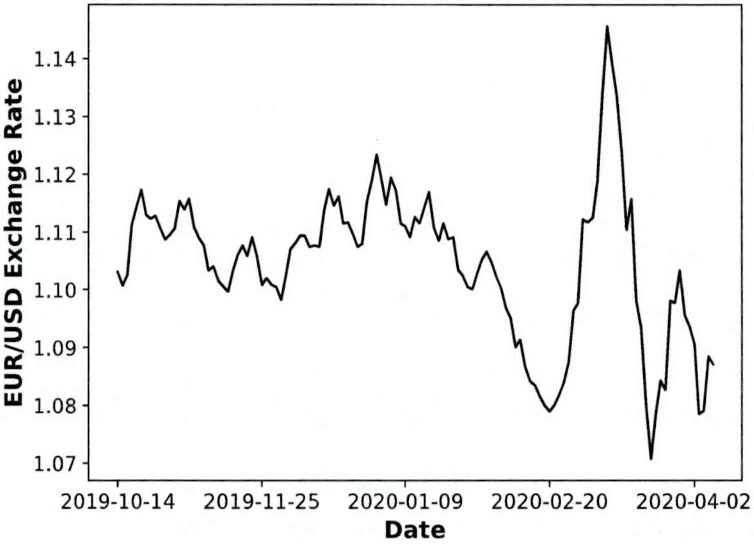

Fig. 1 Exchange rates of EUR to US from 2019/10/14 to 2020/4/9. Last days of February begin to experience higher volatility exchange rates as it can be see in the abrupt peak continued by a sharp decrease

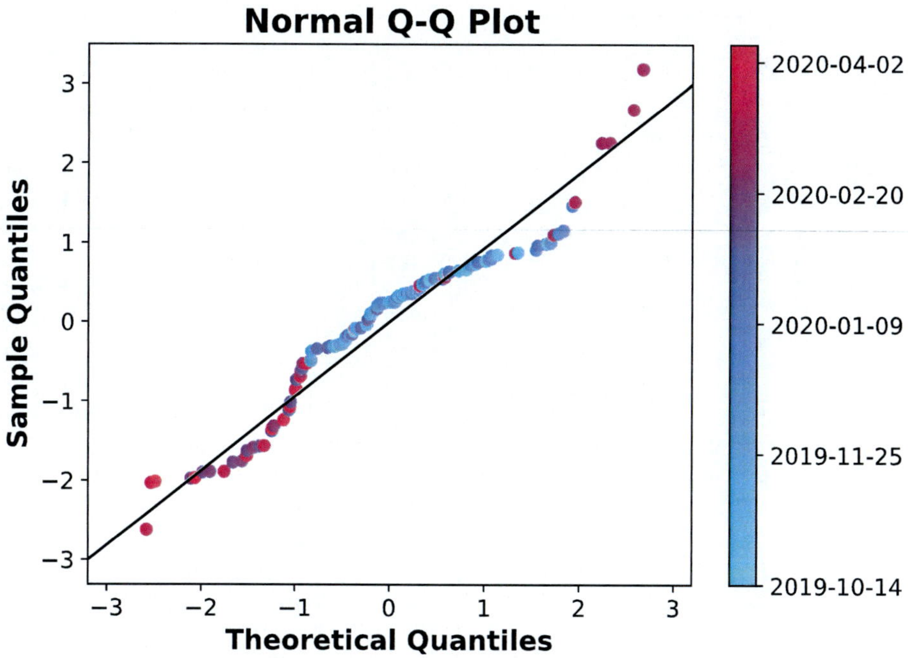

Fig. 2 QQ-plot of exchange rates of EUR to US from 2019/10/14 to 2020/4/9. Dates closer to magenta color are the days in which the COVID-19 has a major impact in the global economy

the dates near the magenta color, i.e. dates that are closer to the pandemic outbreak. We also plotted a histogram of the data in Fig. 3 to potentially identify the strange object further and this was the result. The histogram in Fig. 3 shows that the data is bimodal, meaning that there is a high concentration of two different means. The reason for this outcome could be that the data has the same variance; days of the week. The lower concentration is dealt with COVID-19 impacting world trade hence the exchange rate decreased.

After observing this, we decided to split the data into two in Fig. 4, as the histogram shows there are essentially two different datasets. We split at February 5th. If we compare the two QQ-plots of the split data in Fig. 4, we can see that both are much more normal. While not confidently normal, judging by the outliers and the Shapiro-Wilk values for both of 0.97353 and a p-value of 0.03971, and then 0.9489 with a p-value of 0.02248. This provides another explanation for the object in the original QQ-plot.

The global dollar appreciation can be measured as the Nominal/Real Advanced Foreign Economies Dollar Index [4]. This daily index in Fig. 5 is presented by the Governors of the Federal Reserve System in order to measure the strength of the US Dollar against other currencies that are used in international trades. We shortly present the values of the Index in order to show differences with the exchange rate of EUR/US, which are basically that the COVID-19 pandemic impact negatively

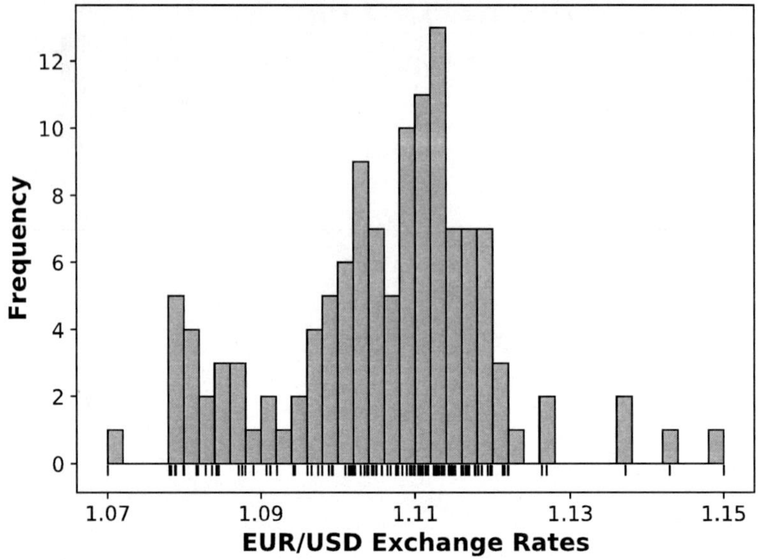

Fig. 3 Distribution of exchange rates of EUR to US from 2019/10/13 to 2020/4/9

the Dollar against EUR which was effectively captured with the Nominal Advanced Foreign Economies Dollar Index. As the EUR/US exchange rate (as depicted in Fig. 1) increases in the last days of February 2020, the Dollary Index loose its strength, decreasing until March 9th, when the global markets begin to fall, causing the liquidation of the volatile assets, such as stocks, in order to change it for safe haven assets as gold or US Dollar.

2 Forecasting EUR/US Exchange Rates

2.1 Finding a Model for Forecasting

In order to predict, we need to forecast the data. In order to do that, we must apply correct models to the data. After careful deliberation and programming, we came across one model for both original data and post-COVID data. We will, however, be focusing more on the post-COVID data as that will be more useful for forecasting in the near future.

Here are the models we found to be most appropriate for the data. Bolded are models we choose to forecast due to lowest AIC, BIC and highest log-likelihood. We found these models by applying autocorrelation function plots, partial autocorrelation function (ACF) plots, sample extended ACF plots, and subset ARMA plots. Then, we eliminated non-significant coefficients to find our very best models.

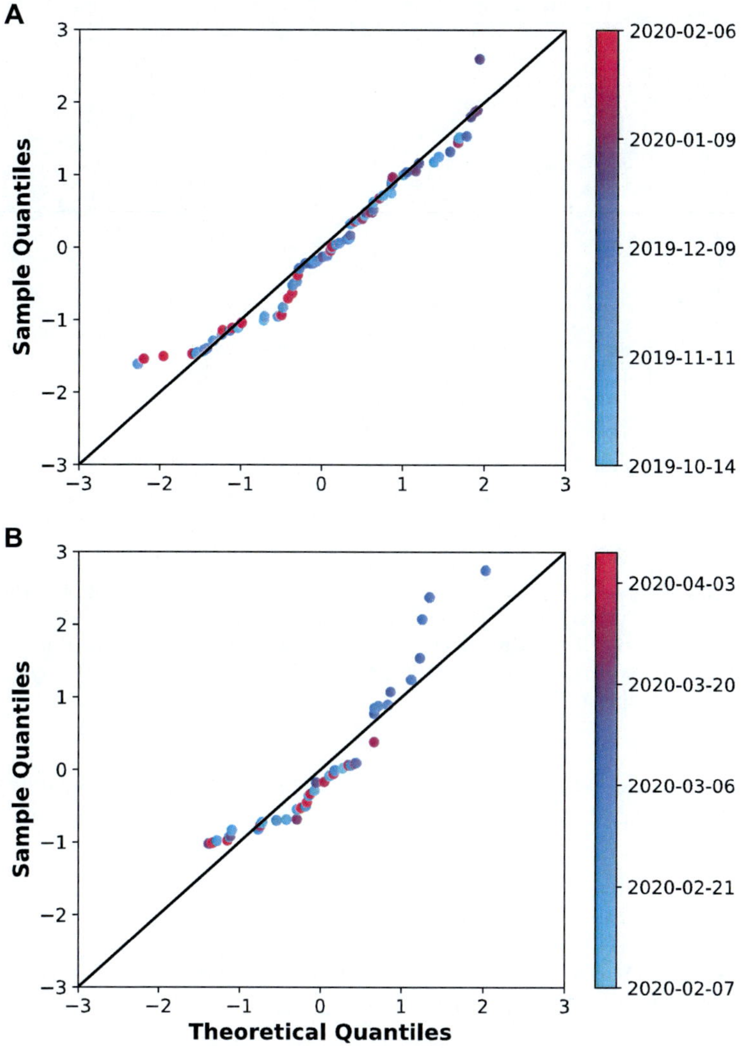

Fig. 4 QQ-plot of exchange rates of EUR to US. (**a**) depicts the data from Q-Q plot of exchange rates from dates 2019/10/14 to 2020/2/6, and (**b**) displays the Q-Q plot for exchange rates from 2020/2/7 to 2020/4/3

In order to find the best model we test several parameters in the ARIMA model ($p, q \in \{0, 1, \ldots, 10\}$ and $d \in \{0, 1, 2\}$) and we choose the ones that has the lowest AIC, BIC and highest log-likelihood.

The meaning of ARIMA is "Autoregressive Integrated Moving Average". AR (AutoRegressive) indicates the strength of the correlation of the data's own previous values on itself. MA (Moving Average) indicates that the regression error is a linear combination of previous error terms. The "I" indicates the difference of its own

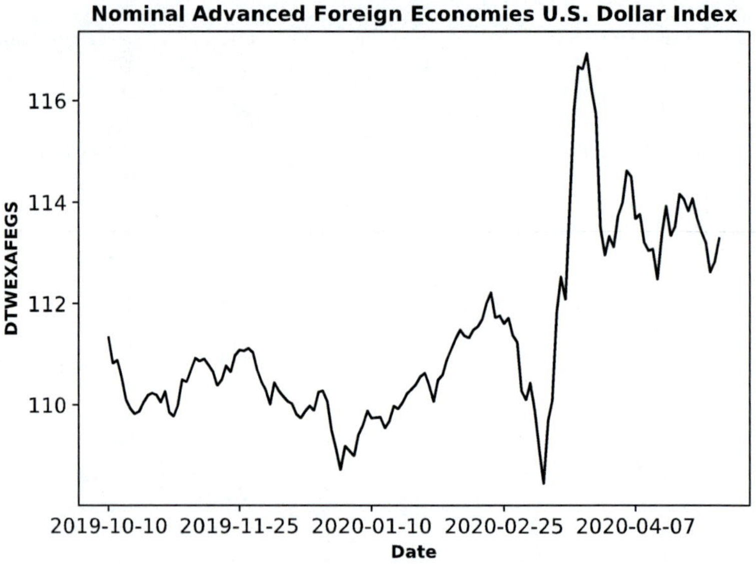

Fig. 5 Federal Reserve Nominal Advanced Foreign Economies U.S. Dollar Index. Data provided by the Board of Governors of the Federal Reserve System (US). This index measures the appreciation of the dollar against other currencies that are used in international trades. In this figure we showed the index from 2019/10/10 to 2020/5/4, i.e. we include months before and after the COVID-19 outbreak

values. If the original data is not stationary, the first difference of the data is needed ($Y_t - Y_{t-1}$), and "I" becomes equal to 1 for the first difference assuming it is then stationary.

2.2 *Forecasting: 34 Days in the Future (Due to the Difference of Observed Data and Current Date)*

As it was seen in the data from the European Central Bank [5] presented in Fig. 1, we can see, the data averages around 1.185–1.19 with a peak of 1.10985 on May 2nd and a low of 1.07444 on April 24th.

If we look at the forecast for the data Fig. 6, it relies heavily on the average of all the data and small rise at the very end. It most certainly does not serve as a good predictor for the data, as it forecasts the data to be above 1.10 on average, which it most certainly is not.

The forecast for the data after the impact of COVID-19 appears to be more accurate in Fig. 7. It continues the trend and stays below 1.10. Most predicted values occur between 1.089 and 1.095. The 95% confidence limits (gray trend lines closest to the middle) naturally increase as time goes on but stay between approximately 1.06 and 1.3, or more closely 1.075 and 1.2, which nearly all 34 real datapoints

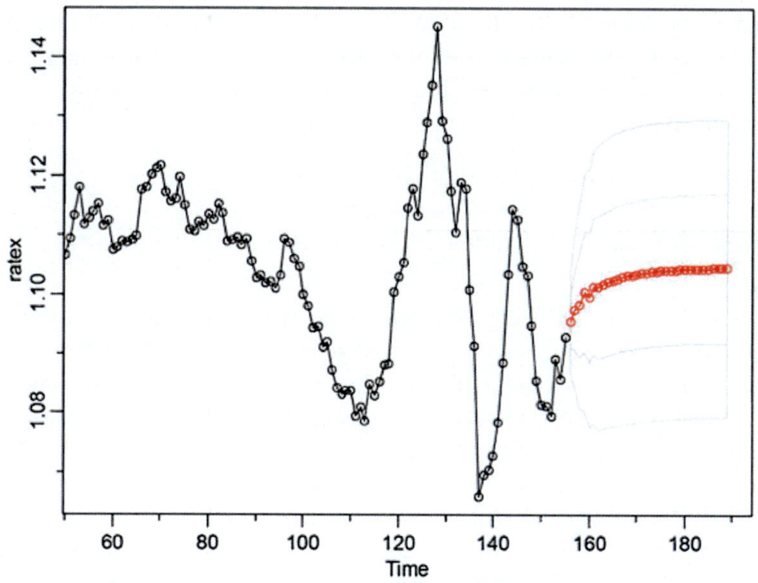

Fig. 6 ARMA(1,7) model with MA 3,4,5,6 missing. The model in this figure is $Y_t = 1.1045 + 0.8677Y_{t-1} + 0.2857e_{t-1} + 0.331e_{t-2} - 0.2403e_{t-7} + e_t$

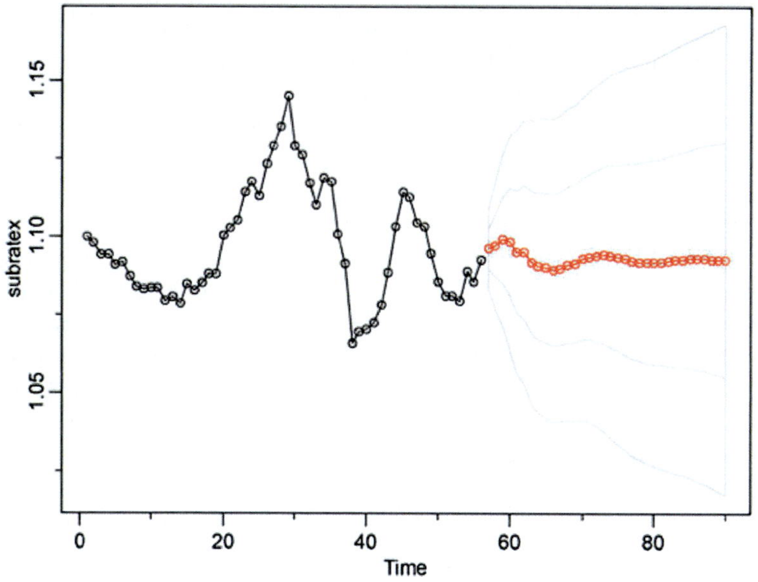

Fig. 7 AR(7) model of the difference with AR 3,5,6 and mean zero. The model in this figure is $\Delta Y_t = 0.1994Y_{t-1} + 0.1918Y_{t-2} - 0.22Y_{t-5} - 0.2633Y_{t-7} + e_t$

apply to. This shows a slight but noticeable strengthen in the USD by nearly 3% after COVID. It also shows a level of consistency to the data, meaning that we can fairly accurately predict that real life EUR to USD exchange rates will most likely be around 1.09.

3 Discussion: What Does This Mean for the Economy?

It is not clear where the abrupt peak of the EUR/USD exchange rate between end of February and beginning of April 2020 originates from. COVID-19 can hardly explain it since the pandemic affected European countries in the same way as it affected the US. An explication could be that the US government declared somewhat earlier than European countries the national emergency concerning the COVID-19 outbreak [6, 7]. In the moment European countries took similar actions as the US government the EUR/USD exchange rate normalized.

Oil Prices are incredibly volatile during harsh economic changes. If we look at oil prices as of recent in Fig. 8, we see a significant dip due to COVID [8]. It fell over 70% in a few weeks, below $20/barrel for the first time since the 1990s. It averaged below $0/barrel during the month of April for the first time ever. This is most likely due to the stay-at-home orders and national hysteria over the pandemic. If we look at historical data, mostly focusing on 2008, we can see there was both a quick yet enormous rise and dip in 2008. It climbed to $143.68/barrel after an increase of 25% in three months. The OPEC blamed the weak US dollar. It also fell below $40/barrel in the next five months. In 2015, EUR to USD exchange rate decreased by nearly 20% (strengthening of USD) and oil dropped severely as well. So we can corroborate the well-known inverse correlation between oil price and USD. Finally, we look at the Federal Funds Rate over time [9] in Fig. 9. The effective federal funds rate is essentially determined by the market. A low rate indicates low demand for overnight loans between depository institutions (banks or credit unions), which in turn means that these institutions hold excess cash, in other words, they fail to place a sufficient volume of loans. Thus, a low FRR indicates low demand for loans, e.g., mortgage loans, loans for business startups, etc. Usually, if the FRR is high the economy is doing well. If it is low, the economy is not doing well. If we look at historical data, we can see it was over 5% in 2006 and 2007 before the recession. It then dipped below 0.10% by December of 2008. In the aftermath of the recession, it began to climb slowly up to 2.40%. But since February 17th, it has dipped to 0.05% and below. This is the lowest it has ever been. We only briefly need to touch on unemployment rate [10] in Fig. 10 as the graph explains itself. Unemployment increases during times of poor economic performance but the current unemployment rate during COVID-19 is unique in that it is astronomically high. It makes the 2008–2010 unemployment seem insignificant.

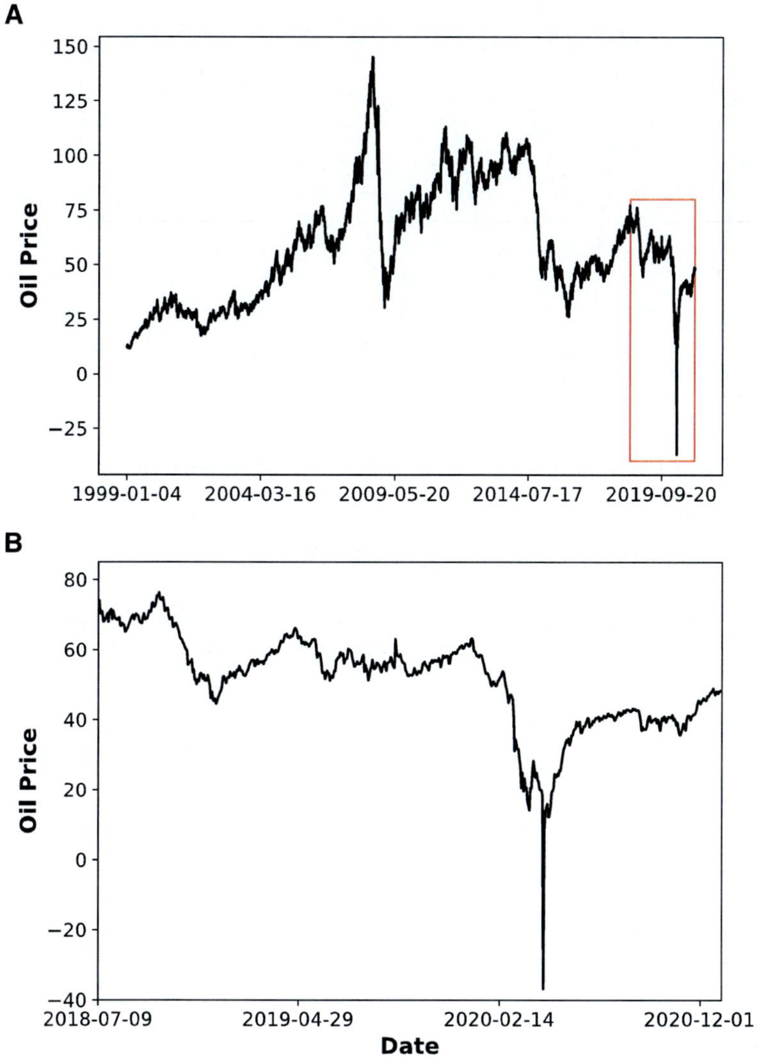

Fig. 8 US Oil price's comparison with historical prices since 1999. Panel (**a**) shows historical data until December 2020, inside this panel the red box is zoomed in panel (**b**) which shows data from November 2019 to April 2020

4 Conclusion

It is evident from this data that USD/EUR exchange rates were bound to fall, and they did, though they seem to currently be stabilizing. It is also obvious, from the data collected, that there is strong evidence to conclude that we are heading towards a major economic recession, assuming we aren't already in one. We claim

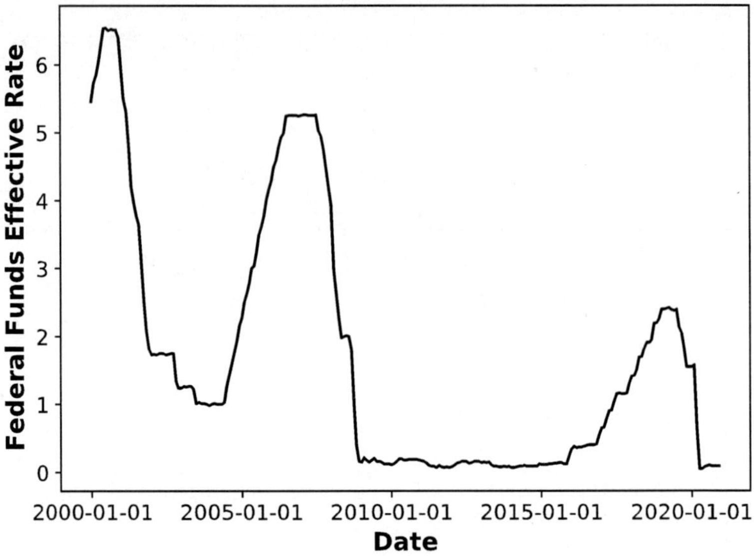

Fig. 9 Federal Funds Rate over time. Before the last two market crashes, i.e. Internet Bubble in the late of 2002 and the financial crisis of 2008, fund rates dropped considerably from previous peak. The same can be seen in months prior the 2020 March black swan

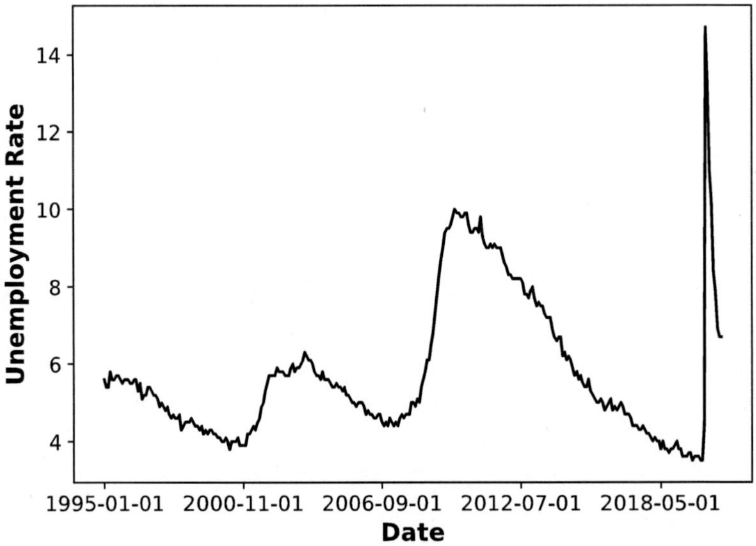

Fig. 10 Unemployment Rate over time. Notice the levels for the unemployment rate, this kind of abrupt peak has never been recorded. Mostly because since 1918 influenza pandemic, we haven't experience a deadly global pandemic with lockdowns that force almost every work in the world to close its doors

that we assumed the models to be correct, since working with real data puts an extra challenges to the simple model fitting. Our approach is more linking to benchmarking of the real data situation. All the figures, except 6 and 7 were made using Python, otherwise was R software.

References

1. Stehlík, M., Helperstorfer, Ch., Hermann, P., Supina, J., Grilo, L.M., Maidana, J.P., Fuders, F., Stehlíková, S.: Financial and risk modelling with semicontinuous covariances. Inform. Sci. **394–395C**, 246–272 (2017). https://doi.org/10.1016/j.ins.2017.02.002
2. Gómez, L.L.S., Torres, S., Kiselak, J., Fuders, F., Ishimura, N., Yoshizawa, Y., Stehlik, M.: Long memory estimation in a non-Gaussian bivariate process. Appl. Math. Comput. **420**, 12687 (2022). https://doi.org/10.1016/j.amc.2021.126871
3. Stehlik, M., Leal, D., Kiselak, J., Leers, J., Strelec, L., Fuders, F.: Stochastic approach to heterogeneity in short-time announcement effects on the Chilean stock market indexes within 2016–2019. Stoch. Anal. Appl. **21**(1) (2023). https://doi.org/10.1080/07362994.2022.2164508
4. Summary Measures of the Foreign Exchange Value of the Dollar. Federal Reserve: https://www.federalreserve.gov/releases/h10/summary/. Last Accessed 10 Jul 2022
5. European Central Bank: https://www.ecb.europa.eu/stats/policy_and_exchange_rates/euro_reference_exchange_rates/html/eurofxref-graph-usd.en.html. Last Accessed 10 Jul 2022
6. Register, F.: Declaring a National Emergency Concerning the Novel Coronavirus Disease (COVID-19) Outbreak (2020). Available at: https://www.federalregister.gov/documents/2020/03/18/2020-05794/declaring-a-national-emergency-concerning-the-novel-coronavirus-disease-covid-19-outbreak. Last Accessed 9 Mar 2023
7. Hirsch, C.: https://www.politico.eu/article/europes-coronavirus-lockdown-measures-compared/ (2020). Available at: https://www.politico.eu/article/europes-coronavirus-lockdown-measures-compared/. Last Accessed 9 Mar 2023
8. FRED Economic Data: Crude Oil Prices: West Texas Intermediate (WTI) - Cushing, Oklahoma (DCOILWTICO). https://fred.stlouisfed.org/series/DCOILWTICO. Last Accessed 10 Jul 2022
9. FRED Economic Data: Federal Funds Effective Rate (FEDFUNDS). https://fred.stlouisfed.org/series/FEDFUNDS#. Last Accessed 10 Jul 2022
10. FRED Economic Data: Unemployment Rate (UNRATE). https://fred.stlouisfed.org/series/UNRATE. Last Accessed 10 Jul 2022

Natural Risk Assessment of Italian Municipalities for Residential Insurance

Selene Perazzini, Giorgio Gnecco, and Fabio Pammolli

Abstract In this work, we propose a catastrophe modeling approach to flood and earthquake risk assessment for residential buildings in Italy. This work aims at supporting governors in the definition of a natural risk management strategy. To detect the critical areas of the territory, we compute expected losses per square meter, per municipality, and per structural typology. Our approach allows us to identify the areas where the exposure strongly affects the risk due to the high inhabited density or the presence of fragile buildings. This information is of major relevance for disaster risk reduction. We find that earthquakes in Italy generate annual expected losses approximately equal to 6234.67 million Euros, while flood expected losses amount to about 875.90 million Euros per year. Although earthquakes produce the highest expected losses at the national level, flood losses per square meter often exceed the corresponding earthquake ones.

Keywords Risk assessment · Catastrophe modeling · Earthquake · Flood · Italy

1 Introduction

The substantial lack of good-quality data on losses hinders the assessment and prediction of the financial cost of natural hazards. In order to overcome this issue,

S. Perazzini (✉)
DMS Statlab, Department of Economics and Management, Brescia, Italy
e-mail: selene.perazzini@unibs.it

G. Gnecco
AXES Research Unit, IMT School for Advanced Studies Lucca, Lucca, Italy
e-mail: giorgio.gnecco@imtlucca.it

F. Pammolli
Department of Management, Economics, and Industrial Engineering, Polytechnic University of Milan, Milan, Italy
e-mail: fabio.pammolli@polimi.it

© The Author(s), under exclusive license to Springer Nature Switzerland AG 2023

C. P. Kitsos et al. (eds.), *Statistical Modelling and Risk Analysis*,
Springer Proceedings in Mathematics & Statistics 430,
https://doi.org/10.1007/978-3-031-39864-3_11

insurers are increasingly relying on catastrophe risk models for premium rating and financial planning [17]. These models compute expected monetary losses by estimating and combining four fundamental components of risk [11]:

- **Hazard** (H): it provides a phenomenon description based on physical measures, usually frequency, severity, and location.
- **Exposure** (E): it identifies the object at risk.
- **Vulnerability** (V): it defines the relationship between hazard and exposure, quantifying the impact of the natural event on the object at risk.
- **Loss** (L): it converts physical damages into monetary losses.

Although this line of research is growing fast, not many models are currently available. In fact, catastrophe models require a large amount of information from different sources. Moreover, they strongly depend on the geographical and urban features of the area on which they have been defined and can hardly be adapted to countries other than those for which they have been produced [14, 20]. In this work, we propose a catastrophe modeling approach to flood and earthquake risk assessment for residential buildings in Italy. The country is highly exposed to the two perils, but the current literature only offers a few analyses. Since earthquakes and floods can be assumed to be independent [6, 21], we assess them separately. We estimate earthquake losses using the model developed in [5], which we extend by applying the most recent seismic risk maps by the Italian National Institute of Geophysics and Volcanology (INGV) for the hazard module and a more accurate representation of the exposure. Then, we propose a model for flood risk assessment.

This work aims at supporting governors in the definition of a natural risk management strategy able to enhance the social and financial resilience of the country. In order to detect the critical areas of the territory, we consider the Italian municipalities and compute the expected losses of the country by aggregation. Moreover, our approach identifies the areas where the exposure strongly shapes the risk profile due to the high inhabited density or the presence of fragile buildings. This information can be useful for urban planning purposes. We find that earthquakes in Italy generate annual expected losses approximately equal to 6234.67 million Euros, while flood expected losses amount to about 875.90 million Euros per year. Although earthquakes produce the highest expected losses at the national level, flood losses per square meter often exceed the corresponding earthquake ones.

The work organizes as follows: Sect. 2 presents the database; Sects. 3 and 4 present the earthquake and flood risk models, respectively; Sect. 5 shows the results; Sect. 6 concludes.

2 Data

Information on the Italian building stock has been collected from three datasets. In particular, the number of buildings per municipality, number of storeys, material, and year of construction have been taken from the "Mappa dei Rischi dei Comuni

Italiani" (MRCI) by the Italian National Institute of Statistics (ISTAT); the average number of apartments per municipality has been extracted from the 2015 census by ISTAT; for the average apartment's square meters we referred to [1].

For the earthquake model, the Peak Ground Acceleration (PGA) from the hazard maps released by INGV [12] and the stratigraphic and topographic amplification factors by [7] have been used. The two pieces of information cover 6404 over 7904 municipalities in Italy. Please note that Sardinia is not included in the analysis, as the region is not exposed to earthquakes, and hazard maps are not available for the area.

As far as floods are concerned, we represented hazard by means of flood frequency and depth, which were fitted on the records from the "Aree Vulnerate Italiane" (AVI) archive released by the Italian National Research Council (CNR) [13]. The AVI database refers to "events" as floods that have lasted for days or weeks and may have hit several municipalities or regions. Overall, the events cover approximately 12000 locations and one century. The database is currently the best representation of floods in the twentieth century in Italy and contains a considerable amount of information. However, it presents a number of limitations due to the complexity of the Italian territory and the different sensitivity and knowledge of the impacts of floods in the territory. In particular, the archive collects all the available information on floods nationwide, most of which was gathered from local press and municipal archives. Many variables in the database are incomplete, descriptive, and difficult to compare among different parts of the country. For this reason, we were forced to restrict our analysis to the events for which the variables we needed were available (i.e., 795 events for flood frequency and 475 events for depth). In particular, flood depth is missing for most of the events in the AVI database and sometimes depth measures are replaced by hydrometric heights. We excluded hydrometric heights and assumed that depth levels reported in the database always correspond to the maximum reached in the area, which is a reasonable hypothesis since records in AVI are largely gathered from local press or compensation claims. Several depth measures are often reported for an event. Among those, we considered the maximum value reported for each event. In order to improve the estimate, we combined information in the AVI archive with the most recent and accurate flood maps. The maps are not available for the Marche region and for some parts of Sardinia. Therefore, the corresponding municipalities are not included in the analysis. Overall, we were able to estimate flood losses for 7772 municipalities.

3 Earthquake Risk Assessment

For earthquake risk assessment we refer to the model developed in [5]. While the basic model is applied with no substantial changes, we improve the accuracy of its estimates in two aspects: (i) we use the latest released data on hazard and fit the PGA probability distribution; (ii) for what concerns exposure, we apply the model to a more detailed real-estate database on residential housing.

Losses per square meter, structural typology j, and municipality c are estimated as:

$$l^s_{j,c} = \frac{1}{K_j}\sum_{k=1}^{K_j}\sum_{LS=1}^{N_{LS_k}} RC_k(LS)\int_0^{+\infty}\left[P_k\left(LS|PGA\right) - P_k\left(LS+1|PGA\right)\right]\mathrm{d}F_c\left(PGA\right) \tag{1}$$

where PGA is the peak ground acceleration, $F_c(PGA) = 1 - \lambda_c(PGA)$ is the cumulative distribution function of PGA for the c-th municipality, and $\lambda_c(PGA)$ is its exceedance probability, modeled in the next paragraph related to hazard. Please note that [5] integrates the PGA on $[0, 2g]$ with g gravity acceleration units, but we extended the domain to $[0, +\infty)$ in order to better model the right tail of the PGA distribution, which is associated with rare and catastrophic events. LS indicates a finite set of "limit states" representing subsequent levels of damage (typically ranging from "no damage" to "collapse"). $P\left(LS|PGA\right)$ is the "fragility curve" and represents the conditional probability that a building will be damaged to a certain limit state, as a consequence of a given PGA.[1] For each structural typology j, the model considers a set of K_j fragility curves, each of which is indexed by k and is defined on N_{LS_k} limit states. Details about the specific set of fragility curves are summarized in [5, Table 3]. At last, $RC_k(LS)$ is a function that quantifies the monetary losses associated with a given limit state LS.

The losses per square meter are multiplied by the municipal exposure and aggregated into municipal seismic losses L^s_c:

$$L^s_c = \sum_{j=1}^{5} l^s_{j,c} \cdot E^s_{j,c} \tag{2}$$

where $E^s_{j,c}$ is the number of square meters covered by buildings of the j-th structural typology in the c-th municipality.

Hazard Seismic hazard is represented by PGA and its annual probability of exceedance. INGV released one seismic map for each of 9 probabilities of exceedance in 50 years [16]. Each PGA measure in the seismic maps is georeferenced to a 0.05-degree grid. We associated each cell of the grid with

[1] In more detail, if one denotes by X the random variable whole realization is the PGA, then one defines $P\left(LS|PGA\right)$ as follows:

$$P\left(LS|PGA\right) = \lim_{\varepsilon\to 0^+}\frac{P\left(LS|PGA-\varepsilon \le X \le PGA+\varepsilon\right)}{P\left(PGA-\varepsilon \le X \le PGA+\varepsilon\right)}.$$

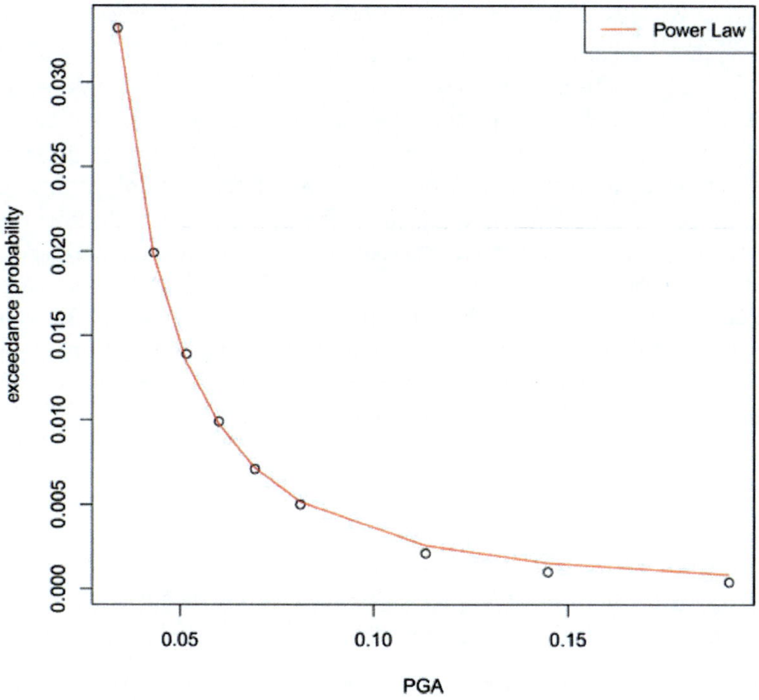

Fig. 1 The plot shows the PGA exceedance probability of a random municipality. The nine points are data collected by INGV, and the red line represents fitting with the power law distribution

the corresponding municipality through reverse geocoding. The process led to the reconstruction of the PGA distribution for over 4600 municipalities. When multiple cells referred to the same municipality their average value was considered. Municipalities for which no grid cell was available were approximated by averaging the neighbors' PGA values. Overall, we were able to capture 7685 municipalities.

The 9 PGA measures available for each grid cell were exploited to extract information about the right tail of $\lambda_c(PGA)$. As one can notice from Fig. 1, these measures do not appear to be uniformly distributed, as assumed in [5]. Therefore, we reconstructed the right tail of the PGA exceedance probability curve for each cell of the grid by fitting such measures and extrapolating outside the range covered by them. More precisely, the measures were fitted by regression, and the best representation was achieved by the power law distribution. At last, the PGA values at the bedrock were multiplied by the stratigraphic S_S and topographic S_T amplification factors in order for the hazard curves to reflect the soil category at the building foundation.

Exposure We refer to the relevant structural typologies identified in [5], which are based on three construction materials: masonry, reinforced concrete, and other. The buildings may have been built in compliance with modern anti-seismic requirements or not. We refer to the construction year in the MRCI database and compare this information to the series of regulations that led to the progressive re-classification of risk-prone areas from 1974[2] to 2003.[3] We define a reinforced concrete or other structure as seismic loaded if built after the laws entered into force in the municipality, and gravity loaded otherwise. Since the database only specifies the time interval (approximately ten years long) in which the building was constructed, we equally distributed the number of buildings constructed among the years in the interval. According to [5], we assumed masonry as seismic loaded only. Summing up, we have 5 structural typologies: masonry, gravity or seismic-loaded reinforced concrete, gravity or seismic-loaded other-type structures.

We compute $E^s_{j,c}$ as

$$E^s_{j,c} = \bar{s}_c \cdot B_{j,c} \cdot \bar{A}_c \tag{3}$$

where $B_{j,c}$ is the number of buildings of type j in c, $\bar{s}_c$ is the average apartments' surface in c, and $\bar{A}_c$ is the average number of apartments per building in c.

Vulnerability Seismic vulnerability is captured by means of fragility curves, that provide the probability of exceeding a certain limit state, given a certain PGA. We apply the selection of fragility curves considered in [5].

Loss The loss component is represented by the function $RC_k(LS)$ that converts structural damages into monetary losses:[4]

$$RC_k(LS) = \left(\frac{LS}{N_{LS_k}}\right) RC \tag{4}$$

where each limit state is represented by a positive integer. We assume that the property value is equal to its reconstruction cost RC (on average 1500 Euros per square meter, assumed to be constant among all the municipalities).

---

[2] Law n. 64, 2 Feb 1974 "Provvedimenti per le costruzioni con particolari prescrizioni per le zone sismiche".

[3] O.P.C.M. 3274 2003 "Primi elementi in materia di criteri generali per la classificazione sismica del territorio nazionale e di normative tecniche per le costruzioni in zona sismica".

[4] We assume a linear relationship between structural damages and monetary losses, likewise in [5]. The reader can refer to that reference for a discussion about the linearity assumption, and about the possibility to generalize it to the nonlinear case.

4 Flood Risk Assessment

In this Section, we propose a model for flood risk assessment. We estimate flood losses per square meter, structural typology j, and municipality c as

$$l_{j,c}^{f} = \frac{RC}{100} \cdot P_c(N_F \geq 1) \cdot \int_{0}^{+\infty} g_j(\delta) f(\delta|flood)\mathrm{d}\delta \tag{5}$$

where δ indicates the depth reached by the flood, g_j represents the depth-percent damage curves for the j-th structural typology, $P_c(N_F \geq 1)$ is the probability of occurrence of at least one flood in one year in the c-th municipality, and $f(\delta|flood)$ is the conditional probability density that a flood reaches a certain depth δ (conditional to the flood occurrence).

We compute the municipal flood losses L_c by aggregation as

$$L_c^f = \sum_{j=1}^{3} l_{j,c}^f \cdot E_{j,c}^f \tag{6}$$

where $E_{j,c}^f$ is the number of square meters covered by buildings of the j-th structural typology in the c-th municipality.

Hazard The hazard module is represented by means of flood frequency $P_c(N_F)$ (where N_F is the number of floods in one year that hit the c-th municipality) and depth density $f(\delta|flood)$. Both the terms were estimated on data from the AVI database as follows. Rather than selecting a given parametrization, a variety of parametric models was considered. Then, the parametric model with the best fit to the data was chosen.

In more detail, in order to represent flood frequency, we fitted the discrete probability density function of the number of floods N_F in a year, $f_{N_F}(N_F)$. In order to capture differences among the municipalities, data were divided into two clusters—A_{P_1} (120 obs.) and A_{P_2} (620 obs.)—on the basis of the hydrological hazard index $P2$ from the flood risk maps (the two clusters refer, respectively, to $0 < P2 < 0.5$, and to $P2 \geq 0.5$). The best fit for the frequency $f_{N_F}^{A_P}$ was obtained by the negative binomial. Indeed, the left plot in Fig. 2 shows that the negative binomial approximates quite well $f_{N_F}^{A_P}$ in each of the two clusters. Despite the curves appearing very similar, they strongly differ in mean (the average number of floods per year is 11.95 in A_{P_1} and 42.58 in A_{P_2}). Instead, the second-best distributions obtained—i.e., the geometric and exponential distributions—turned out not to properly fit the data.

Each flood affects a certain number of municipalities within the cluster A_P. Therefore the probability that c will be flooded at least once in a year is given by $f_{N_F}^{A_P} \cdot \frac{\bar{c}^f}{N_c^{AP}}$ where $\bar{c}^f$ is the average number of flooded municipalities in A_P and

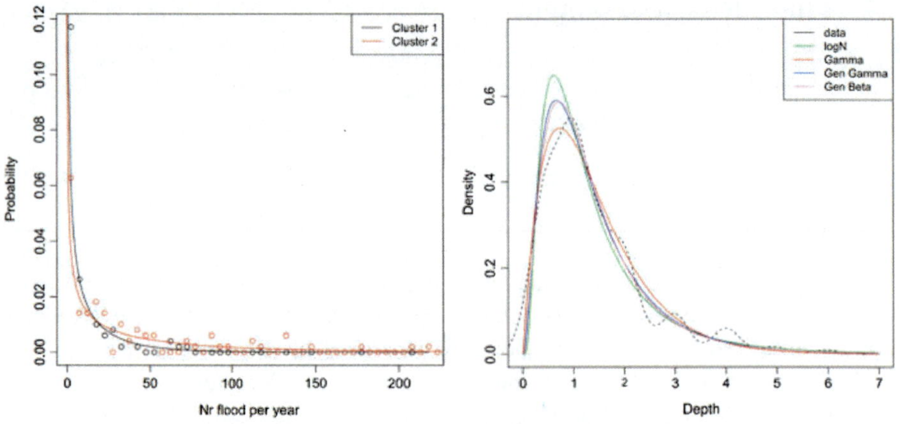

Fig. 2 Left: Flood frequency discrete probability density. Observations (points) are divided into two clusters—records from municipalities with $0 < P2 < 0.5$ (cluster 1) and $P2 \geq 0.5$ (cluster 2)—and fitted with a negative binomial distribution. Right: Depth probability density. The dotted line is the empirical density $f_{\delta|N_F}(\delta|N_F \geq 1)$ (where N_F is the number of floods in a year), and colored lines show the fitting

$N_c^{A_P}$ is the number of municipalities in A_P. During a flood, not all the properties in a municipality get flooded. Therefore, we adjusted the flood probability by the $P3$ index in the flood risk maps[5] indicating the percentage of municipal surface flooded in a 20–50 year probabilistic scenario. Summing up, we compute the probability that a property will be hit by at least one flood in a year as:

$$P_c(N_F \geq 1) = (1 - f_0^{A_P})\frac{\bar{c}^f}{N_c^{A_P}} P3_c. \tag{7}$$

As far as flood depth is concerned, we found no significant difference in depth distribution between the municipalities. As shown in the right plot of Fig. 2, satisfactory fittings of the flood depth are obtained by the generalized Beta, the generalized Gamma, and the Gamma distributions. A Chi-squared goodness of fit test weakly indicates that the generalized Gamma and Beta better fit the distribution, while the likelihood ratio test suggests the opposite. The Gamma was therefore chosen for computational advantages.

Exposure Structural fragility to floods is strongly determined by the number of storeys of the building. We distinguish between three structural typologies: having 1, 2, and 3 or more storeys. The presence or absence of the basement floor also

[5] Index $P3$ is not available for the entire Italian territory: data are missing for part of the Marche and Emilia-Romagna Regions.

significantly affects structural fragility. Since this information is not available, we assumed that half of the buildings have a basement. We represent the exposure as

$$E^f_{j,c} = \bar{s}_c \cdot B_{j,c} \cdot \bar{A}_c \tag{8}$$

where $B_{j,c}$ is the number of buildings with structural typology j within the municipality c.

Vulnerability We used depth-damage curves for the vulnerability component of the model. The most widely adopted curves in the hydraulic literature are the "depth-percent damage curves", which represent the average damage that a building suffers during a flood that reaches a certain depth as a percentage of the building value. These curves are not affected by monetary volatility and are more reliable than the ones expressing damages in absolute values [3].

The depth-damage curves are constructed from historical data and reflect the characteristics of the area on which they have been estimated. Therefore, they tend to be inaccurate when applied to contexts whose urban and territorial features differ too much from the original site [20]. For this reason, we selected six depth-percent damage curves $g_j(\delta)$ from the literature either defined or tested on Italian data [3, 4, 9, 10, 15, 18]. The selected curves represent all or some of the structural typologies and are shown in Fig. 3. Curves were averaged in order to guarantee higher reliability of the results at the national level and fitted by polynomial regression.

Loss Structural damages were converted into monetary losses by means of the factor $\frac{RC}{100}$. Similar to the earthquakes model, we assumed that the property value is equal to its reconstruction cost RC (on average 1500 Euros per square meter, assumed to be constant among all the municipalities).

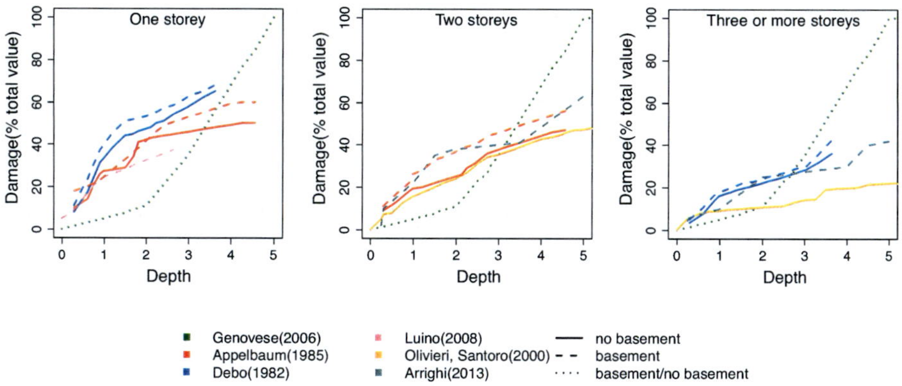

Fig. 3 Depth-percent damage curves for flood risk assessment. Curves are listed per buildings' number of storeys and can refer to dwellings with and/or without a basement

5 Results

Earthquake and flood losses were estimated per municipality and structural typology. Results for earthquakes and floods are shown in Tables 1 and 2 respectively. We found that earthquakes in Italy generate annual expected losses approximately equal to 6234.67 million Euros, while flood expected losses amount to about 875.90 million Euros per year. Although earthquakes produce the highest expected losses at the national level, flood losses per square meter often exceed the corresponding earthquake ones. This happens because of the different extent of the areas exposed to the two perils: while almost all the Italian territory is exposed to earthquakes, floods affect a limited area.

Comparing municipal losses and losses per square meter allows us to capture the different effects of hazard and exposure. In particular, our analysis shows that the highest earthquake losses per square meter are associated with sparsely inhabited municipalities in the central area of the Appennino mountain chain. This result reflects the high probability of earthquake occurrence in the area. The highest municipal expected losses correspond to densely populated cities on the coast. Indeed, the probability that a natural phenomenon will hit these cities is quite low, but their large population densities strongly affect their riskiness. As far as floods are concerned, Northern Italy is the most flood-prone area, and the highest

Table 1 Estimated seismic expected losses. The table shows some descriptive statistics about estimated seismic expected losses per structural typology; in order: maximum expected loss per square meter $l^f_{j,c}$, maximum expected loss at the municipal level $L^s_{j,c}$, and total expected loss L^s_j

j	$\max(l^s_{j,c})$ (Euros)	$\max(L^s_{j,c})$ (Mln Euros)	$L^s_j = \sum_c L^s_{j,c}$ (Mln Euros)
Reinf. conc (gravity)	10.53 Castelbaldo (Padova)	216.79 Roma	2223.61
Reinf. conc (seismic)	3.83 Castelbaldo (Padova)	3.54 Roma	130.70
Other (gravity)	4.03 Castelbaldo (Padova)	7.16 Roma	233.76
Other (seismic)	3.22 Castelbaldo (Padova)	0.43 Roma	30.73
Masonry	12.69 Castelbaldo (Padova)	109.54 Roma	3615.87
Total			6234.67

Table 2 Estimated flood expected losses. The table shows descriptive statistics of flood expected losses per number of storeys. In order: maximum expected loss per square meter $l^f_{j,c}$, maximum expected loss at the municipal level $L^f_{j,c}$, and total expected loss L^f_j

j	$\max(l^f_{j,c})$ (Euros)	$\max(L^f_{j,c})$ (Mln Euros)	$L^f_j = \sum_c L^f_{j,c}$ (Mln Euros)
1 storey	19.61 Vigarano M. (Ferrara)	7.93 S. Michele al T. (Venezia)	105.75
2 storeys	15.16 Vigarano M. (Ferrara)	36.53 Ferrara	536.14
3 storeys	11.56 Vigarano M. (Ferrara)	18.24 Rimini	234.01
Total			875.90

expected losses per square meter are estimated around the Po river and correspond to municipalities in the Emilia-Romagna, Veneto, and Lombardia regions. Most of these municipalities are densely inhabited and are therefore also associated with some of the highest municipal flood losses. In addition to this, high municipal expected losses are also estimated on the northwest coast, in north Sardinia and Rome. Finally, northeast Italy is highly affected by both two hazards, though the effect of floods remains consistently limited with respect to that of earthquakes.

Finally, it is worth observing that the estimates obtained with the earthquake and flood models rely on a set of hypotheses and parameters and are therefore uncertain. Unfortunately, it is often not possible to test the predictive ability of natural risk models by means of the traditional statistical techniques [8], mostly due to the general lack of data on past events. For validation purposes, it is worth noticing that IVASS—the Italian insurance supervisory institute—estimated the average annual loss on residential buildings due to seismic events in Italy to be equal to 4.7 billion Euros.[6] This value is quite close to our findings, according to which the total expected loss is approximately equal to 6 billion Euros. As far as the validation of the flood analysis concerns, the report [2] found that the expected losses due to river floods constitute about 8% of the total annual expected loss generated by both river floods and earthquakes. Our results suggest that this ratio, evaluated considering any flood type, is approximately equal to 12%, and it is therefore in line with the aforementioned report.

6 Conclusion

A natural risk assessment analysis for residential buildings was presented. An earthquake catastrophe risk model was extended in order to improve the accuracy of the estimates, and a model for flood risk assessment was proposed. We were able to estimate losses per square meter, at the municipal and national levels. This information is of main relevance for risk financing, especially for insurers. Particularly, the monetary losses derived in the present work form the basis for the residential insurance models investigated in [19]. In that work, indeed, various models for insurance against natural hazards like earthquakes and floods were developed and applied to the case of Italy, taking as starting point of the analysis the risk assessment model which is presented here (in a much more detailed way as in [19]). Moreover, our analysis allows us to identify the municipalities where the risk is strongly affected by the probability of an event and those where the exposure is the main determinant of risk. This finding is particularly important for policy-makers and can be useful for the definition of effective risk reduction strategies. We found that earthquakes produce the highest expected losses, but floods can severely affect a few municipalities.

[6] See [6], p. 35.

References

1. Agenzia delle Entrate: Gli immobili in Italia. Technical report, Agenzia delle Entrate (2015)
2. ANIA, Carpenter, G.: Danni da eventi sismici e alluvionali al patrimonio abitativo italiano. Technical report, ANIA - Associazione Nazionale fra le Imprese Assicuratrici (2011)
3. Appelbaum, S.J.: Determination of urban flood damage. J. Water Resour. Plan. Manag. **111**(3), 269–283 (1985)
4. Arrighi, C., Brugioni, M., Castelli, F., Franceschini, S., Mazzanti, B.: Urban micro-scale flood risk estimation with parsimonious hydraulic modelling and census data. Nat. Hazards Earth Syst. Sci. **13**(5), 1375–1391 (2013)
5. Asprone, D., Jalayer, F., Simonelli, S., Acconcia, A., Prota, A., Manfredi, G.: Seismic insurance model for the Italian residential building stock. Struct. Saf. **44**, 70–79 (2013)
6. Cesari, R., D'Aurizio, L.: Natural disasters and insurance cover: risk assessment and policy options for Italy. IVASS Working Paper No. 12 (2019)
7. Colombi, M., Crowley, H., Di Capua, G., Peppoloni, S., Borzi, B., Pinho, R., Calvi, G.M.: Mappe di rischio sismico a scala nazionale con dati aggiornati sulla pericolosità sismica di base e locale. Progettazione Sismica (2010)
8. de Moel, H., Jongman, B., Kreibich, H., Merz, B., Penning-Rowsell, E., Ward, P.J.: Flood risk assessments at different spatial scales. Mitig. Adapt. Strat. Glob. Chang. **6**(20), 865–890 (2015)
9. Debo, T.N.: Urban flood damage estimation curves. J. Hydraul. Div. **10**(108), 1059–1069 (1982)
10. Genovese, E.: A methodological approach to land use-based flood damage assessment in urban areas: Prague case study. JRC Report - EUR 22497 (2006)
11. Grossi, P., Kunreuther, H., Windeler, D.: An introduction to catastrophe models and insurance. In: Grossi, P., Kunreuther, H. (eds.) Catastrophe Modeling: A New Approach to Managing Risk. Springer, Berlin (2005)
12. Gruppo di Lavoro MPS: Redazione della mappa di pericolosità sismica prevista dall'Ordinanza PCM 3274 del 20 marzo 2003. Rapporto Conclusivo per il Dipartimento della Protezione Civile (2004).
13. Guzzetti, F., Tonelli, G.: Information system on hydrological and geomorphological catastrophes in Italy (SICI): a tool for managing landslide and flood hazards. Nat. Hazards Earth Syst. Sci. Copernicus Publications on behalf of the European Geosciences Union **4**(2), 212–232, (2004)
14. Hufschmidt, G., Glade, T.: Vulnerability analysis in geomorphic risk assessment. In: Geomorphological Hazards and Disaster Prevention, Cambridge University Press, pp. 233–243 (2010)
15. Luino, F., Cirio, C.G., Biddoccu, M., Agangi, A., Giulietto, W., Godone, F., Nigrelli, G.: Application of a model to the evaluation of flood damage. Geoinformatica **13**, 339–353 (2009)
16. Meletti, C., Montaldo, V.: Stime di pericolosità sismica per diverse probabilità di superamento in 50 anni: valori di ag. Progetto DPC-INGV S1, Deliverable D2 (2007)
17. Mitchell-Wallace, K., Foote, M., Hillier, J., Jones, M.: Natural Catastrophe Risk Management and Modelling: A Practitioner's Guide. Wiley Blackwell, Hoboken (2017)
18. Oliveri, E., Santoro, M.: Estimation of urban structural flood damages: the case study of Palermo. Urban Water **2**, 223–234 (2000)
19. Perazzini, S., Gnecco, G., Pammolli, F.: A public-private insurance model for disaster risk management: an application to Italy. Italian Econ. J. (2022). https://doi.org/10.1007/s40797-022-00210-6
20. Scorzini, A., Frank, E.: Flood damage curves: new insights from the 2010 flood in Veneto, Italy. J. Flood Risk Manage. **10**, 381–392 (2015)
21. Tarvainen, T., Jarva, J., Greiving, S.: Spatial pattern of hazards and hazard interactions in Europe. Geol. Surv. Finland **42**, 83–91 (2006)

Variable Selection in Binary Logistic Regression for Modelling Bankruptcy Risk

Francesca Pierri

Abstract One of the most fascinating areas of study in the current economic and financial world is the forecasting of credit risk and the ability to predict a company's insolvency. Meanwhile, one major challenge in constructing predictive failure models is variable selection. Standard selection methods exist alongside new approaches. In addition, the huge availability of data often implies limitations due to processing time and new high-performance procedures provide tools that can take advantage of parallel processing. In the present paper, different variable selection techniques were explored in the context of applying logistic regression for binary data to a balanced data set including only firms active or in bankruptcy. Models deriving from stepwise selection, the Least Absolute Shrinkage and Selection Operator (LASSO) and an unsupervised method, based on the maximum data variance explained, were compared. Then a non-parametric approach was considered and the selection of variables coming from a single decision tree and a forest of trees is compared and discussed.

Keywords Variable selection · LASSO · Stepwise · Unsupervised methods · Decision trees · Logistic · Unbalanced data

1 Introduction

From 2005 onwards, credit risk forecasting and bankruptcy prediction have become among the most important and interesting topics in the modern economic and financial field. However, quantitative methods have long been applied for predicting the bankruptcy event. First, Beaver in 1966 [5] applied discriminant analysis, then Altman [1] in 1968 developed the well-known Z score. Later on, Ohlson [28] in 1980 used logistic regression which has became the most applied model

F. Pierri (✉)
Department of Economics, University of Perugia, Perugia, Italy
e-mail: francesca.pierri@unipg.it; http://www.stat.unipg.it/~frc/

© The Author(s), under exclusive license to Springer Nature Switzerland AG 2023
C. P. Kitsos et al. (eds.), *Statistical Modelling and Risk Analysis*,
Springer Proceedings in Mathematics & Statistics 430,
https://doi.org/10.1007/978-3-031-39864-3_12

in the credit scoring field. Subsequently, in 1992 Narain [27] approached the problem via survival analysis, examining the timing of failure instead of simply considering whether or not an event occurred within a fixed interval of time; since then, Cox's semi-parametric proportional hazard model and its extensions have been extensively proposed and adopted in economic, banking and financial fields [4, 6, 9, 20, 30, 38, 39].

However, whichever model is applied, one major challenge in constructing predictive failure models, as has been widely stated in the literature [2, 3, 7, 8, 15–19], is the effective selection of the most relevant variables from among those that have been collected because of their perceived importance or widespread use.

Besides the problem of correlations between variables that may affect the discriminant ability of a risk model [24], a crucial point remains the procedure chosen for making the selection [13, 45]. Beyond the traditional methods such as backward, forward and stepwise selection, and the use of criteria such as the Akaike Information Criterion (AIC) and the Bayesian Information Criterion (BIC), new approaches known as penalty driven methods (Least Absolute Shrinkage and Selection Operator (LASSO), Smoothly Clipped Absolute Deviation (SCAD) or bridge estimator) [21, 41–44] and machine learning techniques (decision trees and neural networks) [11, 23, 25, 40] have become prominent. Moreover, the increased availability of high-dimensional data, which may impose limitations due to processing time, has led to the development of new high-performance procedures employing tools that can take advantage of parallel processing [37].

In the present paper, based on an application to economic data, we try to provide an answer to the following research questions: (1) do different variable selection methods among standard, modern and those taking advantage of parallel processing, lead to the same choice of variables; (2) which method is better for predicting the future state of a firm.

The paper is structured in the following way: Sect. 2 presents the methodology that will be applied; Sect. 3 gives a brief description of the data; results of the analysis are shown in Sect. 4; and Sect. 5 presents the conclusions of the investigation.

2 Methodology and Study Design

The primary purpose of this paper is to apply different techniques in order to select significant variables for predictive purposes, applying as quantitative method the binary logistic regression model. While acknowledging that different causes may lead to the end of a firm's life, that alternative variables may influence these various events, and that the same variables may even have opposite effects (see [10] and [31]), a single adverse event— bankruptcy —was studied. The problem of overestimating the intercept coefficient in the logistic model [22] due to the relative lack of data on rare events, was overcome by applying one of the available solutions that we have previously applied in statistical analysis [32]. Thus a balanced data set was built by randomly selecting for each bankrupt firm four controls (firms that did not fail). Training and holdout samples were built to develop and test the models, respectively. The variables selected as relevant by each method were used as explanatory variables in a logistic model. The Wald test was applied to test whether a candidate variable should be included in the model, with the p-value cutoff set at 0.05. Each model's adequacy and predictive capability were tested, through the holdout sample, measuring the Area Under the Receiver Operating Characteristic (ROC) Curve (AUC).

Three parametric (forward-stepwise, LASSO, Maximum Data Variance (MDV) explained [36]) and two non-parametric methods (single and forest decision tree) were applied and compared, taking into account the number of selected variables and the AUC value in the holdout sample.

Focusing attention on SAS® software, which provides both standard and high performance (HP) procedures running in either single-machine mode or distributed mode, the following procedures were called upon: LOGISTIC [33] to apply the forward stepwise selection and to run and test all the logistic models; GLMSELECT [34], specifying the logit link, to perform the LASSO selection following the Efron et al. implementation [14]; HPREDUCE [37] to identify variables that jointly explain the maximum amount of data variance; and HPSPLIT [35] and HPFOREST [37] to build a single tree and a forest of trees, respectively.

3 Data Description

The data used in this study were extracted from Orbis [29], a global company database compiled by Bureau Van Dijk, one of the major publishers of business information. Orbis combines private company data with software for searching and analysing over 400 million companies.

The sample employed in the present analyses consists of 37,875 Italian firms operating in the manufacturing sector from 2000 to 2018. For each firm, the financial data for the last available year, its legal form, current legal status and geographical location were extracted. Following the classification of company status available in the Orbis database, three main categories of firms' inactivity were identified: closure, liquidation and bankruptcy (Table 1). As indicated earlier in the Introduction, only one of the adverse events, bankruptcy, was taken into account and, due to its rarity (8.74%), a balanced data set was built by randomly choosing four controls (active firms) for each event (bankrupt firm). The data obtained in this way (16,560 observations) were then split at random into training (80% of the total sample, 13,095 observations) and holdout samples (20% of the total sample, 3465 observations) in order to develop and test the models on independent samples.

Table 1 Firms' distribution by status

Status	N	%
Active	34, 046	89.89
Closed	43	0.11
Winding-up	474	1.25
Bankruptcy	3312	8.74
Total	37, 875	100

Table 2 Distribution of firms in the training set, by geographical area

		North West	North East	Centre	South	Insular	Total
Active	N	4305	3316	1728	853	274	10,476
	%	32.88	25.32	13.20	6.51	2.09	80.00
Bankruptcy	N	962	729	489	332	107	2619
	%	7.35	5.57	3.73	2.54	0.82	20.00
	Column %	18.3	18.0	22.1	28.0	28.0	
Total	N	5267	4045	2217	1185	381	13,095
	%	40.22	30.89	16.93	9.05	2.91	100.00

Table 3 Distribution of firms in the training set, by legal form (LC = limited company)

		Partnerships	PrivateLC	PublicLC	Total
Active	N	214	8562	1700	10,476
	%	1.63	65.38	12.98	80.00
Bankruptcy	N	43	2290	286	2619
	%	0.33	17.49	2.18	20.00
	Column %	16.7	21.1	14.4	
Total	N	257	10,852	1986	13,095
	%	1.96	82.87	15.17	100.00

The distribution of firms in the training data set by geographical area (Table 2) shows an increasing percentage of defaulting firms going from the North (18%) to the South (28%). Moreover, private limited companies (21%) seem to be more prone to the adverse event (Table 3).

For each firm indexes or ratios representative of its economic and financial situation were built, taking into account both their perceived importance and widespread use in the literature [1, 5, 12, 26] and the information availability required for the calculation. Correlation problems were solved by including only one of the ratios among those with correlation higher than 0.70. Finally, besides the firm's age, geographical area and legal form, 37 indexes were used (Table 4), including liquidity and solvency ratios, profitability and operating efficiency ratios.

Table 4 Indexes evaluated as potential predictors of the bankruptcy event

ID	Formula	ID	Formula
ind001	ln (EBITDA)	ind079	Quick Assets/Sales
ind004	Operating Revenue/Inventories	ind080	Quick Assets/Total Assets
ind007	Cash flow/Current Liabilities	ind083	Profit (Loss) for period/Shareholders' Funds
ind011	Cash flow/Shareholders' Funds	ind084	EBIT/Shareholders' Funds
ind020	Ln(Total Assets)	ind085	Profit (Loss) for period/Operating Revenue
ind021	(Creditors/Operating Revenue)*360	ind087	Sales/Cash flow
ind031	Current Assets/Current Liabilities	ind088	Sales/Current Assets
ind033	Debtors/Operating Revenue	ind089	Sales/EBIT
ind042	Shareholders' Funds/Total Assets	ind090	Sales/Equity ratio
ind044	Equity/Fixed Assets	ind092	Operating Revenue/Total Assets
ind050	Inventory/Sales	ind093	Sales/Working Capital
ind052	Inventory/Working Capital	ind094	Shareholders' Funds/Capital
ind055	Long Term Debts/Sales	ind104	Sales/Shareholders' Funds
ind056	Long Term Debts/Net Capital	ind105	Working Capital
ind058	Non Current Liabilities/Total Assets	ind116	EBIT/Interest paid
ind060	(Long Term Debt + Loans)/Total Assets	ind117	Long Term Debts/Equity
ind063	Net Income/Cash flow	ind124	Debtors/Current Assets
ind065	Net Income/Fixed Assets	ind132	Equity/Sales
ind072	Non-Current Liabilities/Sales		

Table 5 Variable selection comparison among stepwise, LASSO and maximum data variance explained methods

Variables	Stepwise	LASSO	MDV
N. selected	21	19	13
% In common	61.90	68.42	100
AUC training	0.9081	0.906	0.9040
AUC holdout	0.8908	0.8921	0.8903

4 Results

4.1 Stepwise, LASSO and Maximum Data Variance Selection Methods

The variable selection comparison between the stepwise, LASSO and maximum data variance (MDV) explained techniques, shows good performance of all three methods. Although the best performance in the holdout sample was given by the LASSO (AUC = 0.8921), AUC values under the other methods were extremely close (Table 5). The MDV method selected the smallest number of indexes (13), which in turn are also identified by the other two techniques. As shown in Table 5 the three approaches agree on the selection of more than 60% of the variables.

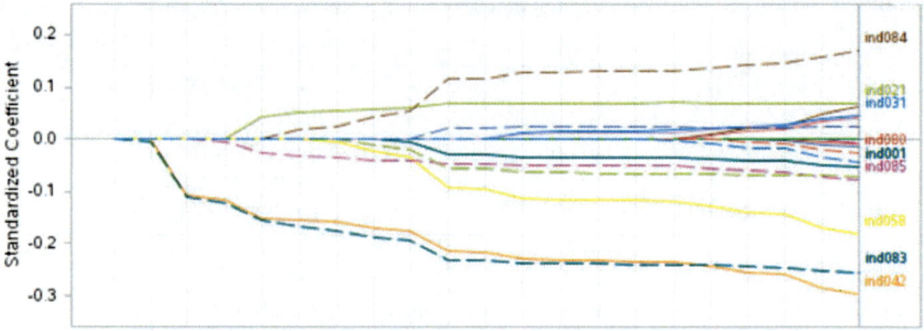

Fig. 1 Coefficient progression for response variable: output from GLMSELECT procedure

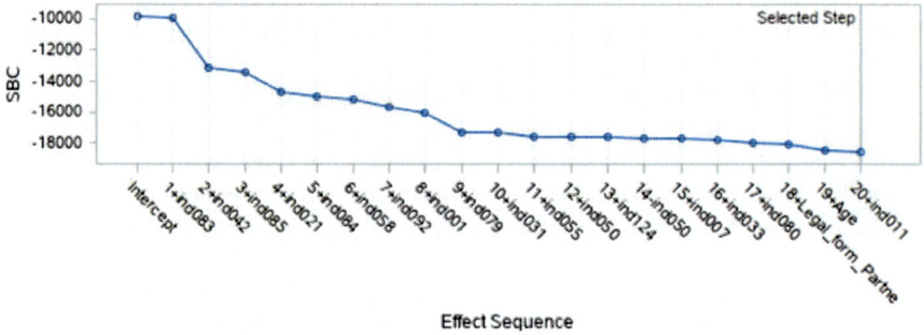

Fig. 2 Effect Sequence: output from GLMSELECT procedure

The LASSO output results from the GLMSELECT procedure include detailed graphs as an aid to interpretation. Figure 1 shows the coefficient progression for the response variable: the names of the most important indexes affecting bankruptcy appear on the right-hand side, with those above the zero line increasing the probability of the event under study when their value increases and those below the zero line decreasing it. Coefficients corresponding to effects that are not in the selected model at a step are zero and hence not observable. Figure 2, complementary to the previous graph, shows how the average square error used to choose among the examined models progresses. The initial model includes only one index (ind0042), then a second one (ind0085) is added and so on (Fig. 2). The procedure stops at the 20th step.

4.2 Single and Forest of Trees Methods

The two non-parametric approaches showed very similar results. The single tree and the forest of trees had in common 12 indexes, that is, respectively, 75% and 80% of the variables selected. Their performances in the holdout sample

Table 6 Variable selection comparison between the two non-parametric approaches

Variable	Single tree	Forest of trees
N. selected	16	15
% In common	75.00	80.00
AUC training	0.9061	0.9037
AUC holdout	0.8892	0.8888

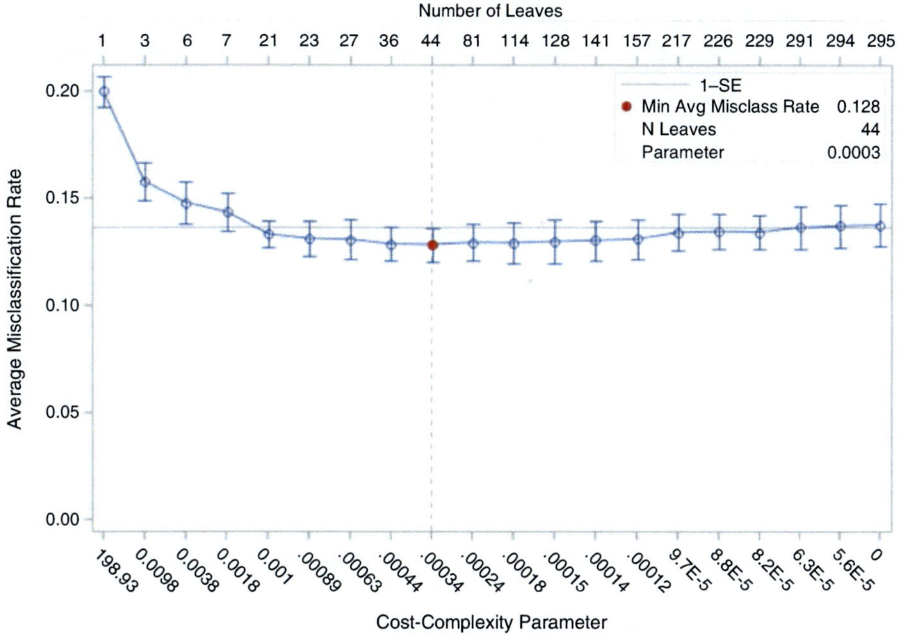

Fig. 3 Cost complexity analysis using cross-validation: PROC HPSPLIT output

were virtually identical (Table 6). HPSPLIT plots provide a tool for selecting the parameters that result in the smallest estimated Average Square Error (Fig. 3) and a classification tree (Fig. 4) that uses colours to aid understanding of where the higher percentage of active firms is found: blue for bankruptcy, and pink for active.

In Fig. 5 the subtree starting at node 0 shows important details regarding the indexes' values, that is, the cut-off at which they cause the separation into new leaves.

4.3 Comparison Between the Best Method of Each Group

Even though all the methods applied in this context lead to very similar results, the best of each group was selected (LASSO and single tree methods) with the aim of making a more detailed comparison among a parametric and non

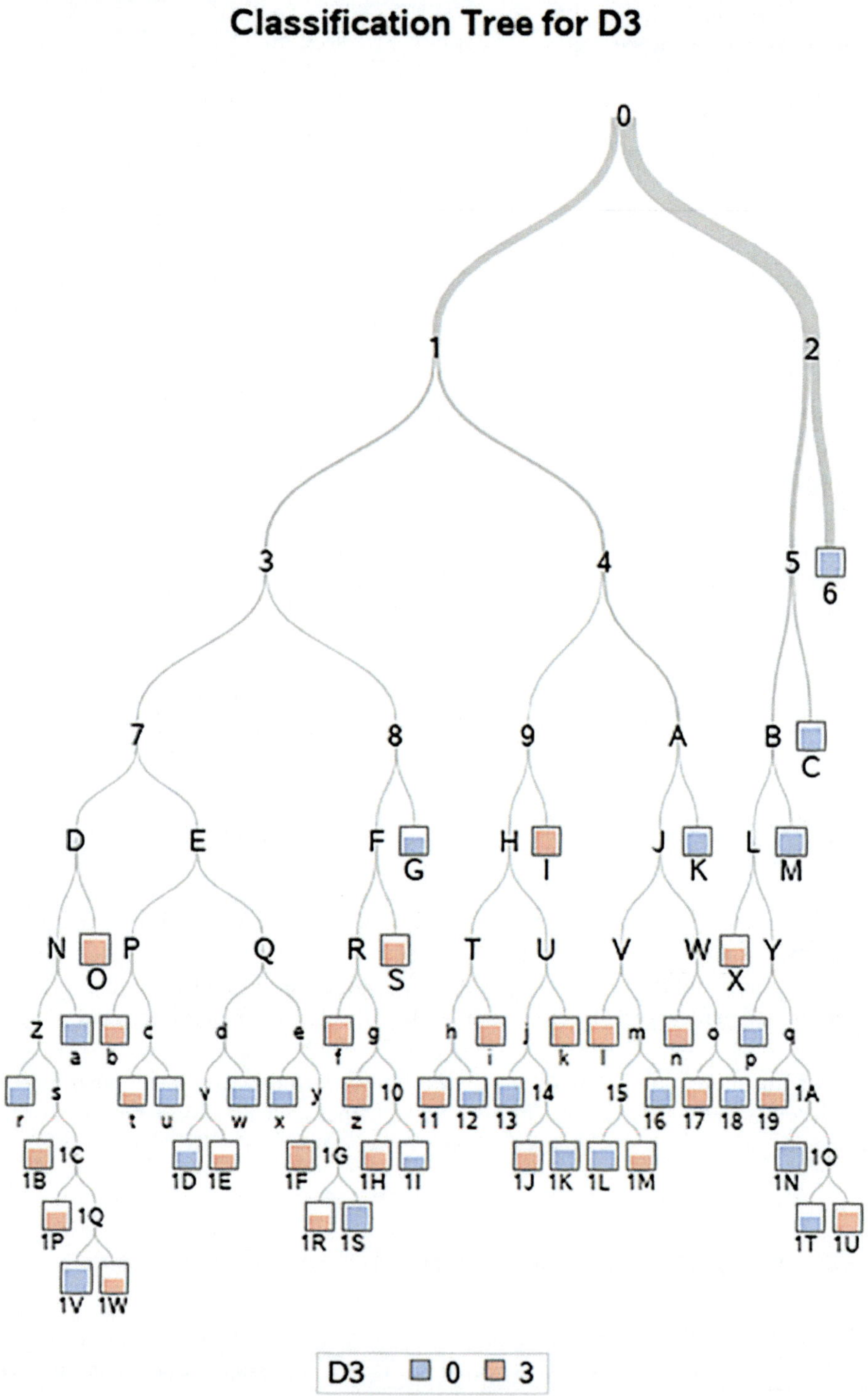

Fig. 4 Classification tree: PROC HPSPLIT output

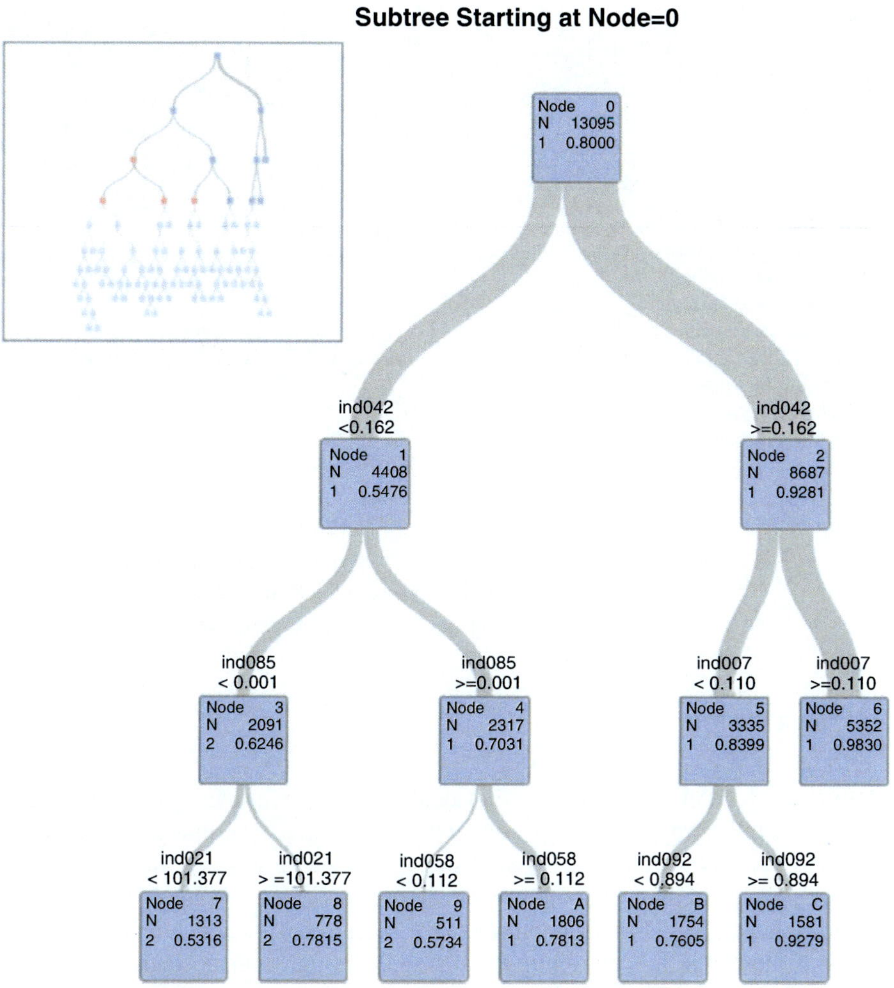

Fig. 5 Subtree starting at node 0: PROC HPSPLIT output

parametric technique (Table 7). The variable selection comparison, on the basis of the AUC value, showed a slight predominance of the first one, however, the difference was extremely small (0.891 against 0.8892). LASSO selected a slightly greater number of variables as predictors, most of which (14) were in common with the single tree method (73.68%). Table 8 shows the ratios that they had in common.

Table 7 Variable selection comparison between the best method in each group

Variable	LASSO	Single tree
N. selected	19	16
% In common	73.68	87.50
AUC training	0.9060	0.9061
AUC holdout	0.8921	0.8892

Table 8 Predictive variables in common between LASSO and single tree methods, in addition to Age and Legal Form. Increased values of variables above and below the horizontal line raise and reduce, respectively, the probability of bankruptcy

ID	Formula
ind021	(Creditors/Operating Revenue)*360
ind031	Current Assets/Current Liabilities
ind033	Debtors/Operating Revenue
ind060	(Long Term Debt + Loans)/Total Assets
ind084	EBIT/Shareholders' Funds
ind001	ln (EBITDA)
ind042	Shareholders' Funds/Total Assets
ind058	Non Current Liabilities/Total Assets
ind083	Profit (Loss) for period/Shareholders' Funds
ind085	Profit (Loss) for period/Operating Revenue
ind092	Operating Revenue/Total Assets
ind124	Debtors/Current Assets

5 Discussion

Variable selection techniques were evaluated within two main groups of methods and then the best of each group were compared further. The first group considered the standard and widely used forward stepwise selection method, the LASSO technique, and a procedure that conducts a variance analysis and reduces dimensionality by selecting the variables that contribute the most to the overall variance of the data. Among these, the models refitted and tested through logistic regression showed very stable results. The AUC values in the holdout sample were very close, with differences only in the third decimal point. The selection was most parsimonious using the third method which discarded variables that are included by both the stepwise and LASSO methods (Table 5), but the AUC value was slightly higher.

The non-parametric approach showed very slight differences between the single tree and the forest methods. Again the differences lay in the third decimal places of the AUC (in the holdout sample) and the number of selected variables was almost the same, with most of these in common.

The final comparison between LASSO and single tree selection methods highlighted that these different techniques led to models with high and stable predictive performance in the holdout sample, with a preference towards the first method for its slightly higher AUC value (0.8921 against 0.8892) and for its computational performance in terms of processing time (0.91 vs. 25.16 seconds). Moreover, the LASSO and single tree approaches selected almost the same predictive variables with a smaller number in the second. In particular both gave particular relevance to variable ind042 reflecting the ratio of Shareholders' Funds to Total Assets: both LASSO and single tree selected it first, on the basis of the average square error and variable importance. This confirms the protection from bankruptcy provided by strong corporate capital structure, while the credit situation (ind021) and debt exposure (ind060) may play an opposite rule [31].

The SAS software procedures used (GLMSELECT and HPSPLIT) both provide very intuitive graphs although perhaps the LASSO ones seem easier to interpret for a wider and non-technical audience. On the other hand HPSPLIT is a high performance procedure that runs in either single-machine mode or distributed mode and can therefore take advantage of parallel processing.

Uniformity in the predictive capability of these selection methods may have been affected by data dimensionality, therefore in the future the same procedures will be applied to a smaller data set. Future developments will also include the extension to multinomial logistic analysis.

References

1. Altman, E.I.: Financial ratios, discriminant analysis and the prediction of corporate bankruptcy. J. Finance **23**(4), 589–609 (1968)
2. Amendola, A., Restaino, M., Sensini, L.: Variable selection in default risk models. J. Risk Model Validation **5**(1), 3 (2011)
3. Austin, P.C., Tu, J.V.: Automated variable selection methods for logistic regression produced unstable models for predicting acute myocardial infarction mortality. J. Clin. Epidemiol. **57**(11), 1138–1146 (2004)
4. Banasik, J., Crook, J.N., Thomas, L.C.: Not if but when will borrowers default. J. Oper. Res. Soc. **50**(12), 1185–1190 (1999)
5. Beaver, W.H.: Financial ratios as predictors of failure. Journal of Account. Res. **4**, 71–111 (1966)
6. Bonini, S., Caivano, G.: The survival analysis approach in Basel II credit risk management: modeling danger rates in the loss given default parameter. J. Credit Risk **9**(1), 101–118 (2013)
7. Bunea, F.: Honest variable selection in linear and logistic regression models via $\ell 1$ and $\ell 1+ \ell 2$ penalization. Electron. J. Stat. **2**, 1153–1194 (2008)
8. Bursac, Z., Gauss, C.H., Williams, D.K., Hosmer, D.W.: Purposeful selection of variables in logistic regression. Source Code Biol. Med. **3**(1), 1–8 (2008)
9. Cao, R., Vilar, J.M., Devia, A., Veraverbeke, N., Boucher, J.P., Beran, J.: Modelling consumer credit risk via survival analysis. SORT Stat. Oper. Res. Trans. **33**(1), 31–47 (2009)
10. Caroni, C., Pierri, F.: Different causes of closure of small business enterprises: alternative models for competing risks survival analysis. Electron. J. Appl. Stat. Anal. **13**(1), 211–228 (2020)
11. Dreiseitl, S., Ohno-Machado, L.: Logistic regression and artificial neural network classification models: a methodology review. J. Biomed. Inform. **35**(5–6), 352–359 (2002)
12. Du Jardin, P.: Predicting bankruptcy using neural networks and other classification methods: the influence of variable selection techniques on model accuracy. Neurocomputing **73**(10), 2047–2060 (2010). https://doi.org/10.1016/j.neucom.2009.11.034, https://www.sciencedirect.com/science/article/pii/S0925231210001098, subspace Learning/Selected papers from the European Symposium on Time Series Prediction
13. Du Jardin, P.: The influence of variable selection methods on the accuracy of bankruptcy prediction models. Bank. Mark. Invest. **116**, 20–39 (2012)

14. Efron, B., Hastie, T., Johnstone, I., Tibshirani, R.: Least angle regression. Ann. Stat. **32**(2), 407–499 (2004)
15. Fan, J., Li, R.: Variable selection for Cox's proportional hazards model and frailty model. Ann. Stat. **30**(1), 74–99 (2002)
16. Fan, J., Li, G., Li, R.: An overview on variable selection for survival analysis. In: Contemporary Multivariate Analysis and Design of Experiments: In Celebration of Professor Kai-Tai Fang's 65th Birthday, pp. 315–336 (2005)
17. Fu, Z., Parikh, C.R., Zhou, B.: Penalized variable selection in competing risks regression. Lifetime Data Anal. **23**(3), 353–376 (2017)
18. Ghosh, K., Ramteke, M., Srinivasan, R.: Optimal variable selection for effective statistical process monitoring. Comput. Chem. Eng. **60**, 260–276 (2014)
19. He, Z., Tu, W., Wang, S., Fu, H., Yu, Z.: Simultaneous variable selection for joint models of longitudinal and survival outcomes. Biometrics **71**(1), 178–187 (2015)
20. Kiefer, N.M.: Economic duration data and hazard functions. J. Econ. Literature **26**(2), 646–679 (1988)
21. Kim, J., Sohn, I., Jung, S.H., Kim, S., Park, C.: Analysis of survival data with group lasso. Commun. Stat. Simul. Comput. **41**(9), 1593–1605 (2012)
22. King, G., Zeng, L.: Logistic regression in rare events data. Political Anal. **9**(2), 137–163 (2001)
23. Kumar, A., Rao, V.R., Soni, H.: An empirical comparison of neural network and logistic regression models. Mark. Lett. **6**(4), 251–263 (1995)
24. Kundu, S., Mazumdar, M., Ferket, B.: Impact of correlation of predictors on discrimination of risk models in development and external populations. BMC Med. Res. Methodol. **17**(1), 1–9 (2017)
25. Meier, L., Van De Geer, S., Bühlmann, P.: The group lasso for logistic regression. J. R. Stat. Soc. Ser. B (Stat. Methodol.) **70**(1), 53–71 (2008)
26. Mossman, C.E., Bell, G.G., Swartz, L.M., Turtle, H.: An empirical comparison of bankruptcy models. Financial Rev. **33**(2), 35–54 (1998)
27. Narain, B.: Survival analysis and the credit granting decision. In: Thomas, L.C., Crook, J.N., Edelman, D.B. (eds.), Credit Scoring and Credit Control, pp. 109–122. Oxford Univeristy Press (1992)
28. Ohlson, J.A.: Financial ratios and the probabilistic prediction of bankruptcy. J. Account. Res. **18**(1), 109–131 (1980)
29. Orbis: Orbis. Bureau van Dijk, https://orbis.bvdinfo.com/. Accessed June 2020
30. Pierri, F., Caroni, C.: Bankruptcy prediction by survival models based on current and lagged values of time-varying financial data. Commun. Stat. Case Stud. Data Anal. Appl. **3**(3–4), 62–70 (2017)
31. Pierri, F., Caroni, C.: Analysing the risk of bankruptcy of firms: survival analysis, competing risks and multistate models. In: Demography of Population Health, Aging and Health Expenditures, pp. 385–394. Springer (2020)
32. Pierri, F., Stanghellini, E., Bistoni, N.: Risk analysis and retrospective unbalanced data. Revstat-Stat. J. **14**(2), 157–169 (2016)
33. SAS: SAS/STAT® 9.22 User's Guide. https://support.sas.com/documentation/cdl/en/statug/63347/HTML/default/viewer.htm#logistic_toc.htm. Accessed 19 Nov 2022
34. SAS: SAS/STAT® 9.22 User's Guide. https://support.sas.com/documentation/cdl/en/statug/63347/HTML/default/viewer.htm#glmselect_toc.htm. Accessed 19 Nov 2022
35. SAS: SAS/STAT® 9.22 User's Guide. https://support.sas.com/documentation/cdl/en/statug/68162/HTML/default/viewer.htm#statug_hpsplit_overview.htm. Accessed 19 Nov 2022
36. SAS: SAS® Enterprise Miner™: High-Performance Procedures. https://documentation.sas.com/doc/en/emhpprcref/14.2/emhpprcref_hpreduce_details01.htm (2016). Accessed 19 Nov 2022
37. SAS Institute Inc., Cary, NC: SAS® Enterprise Miner™ 15.2: High-Performance Procedures, last updated: August 18, 2022
38. Shumway, T.: Forecasting bankruptcy more accurately: a simple hazard model. J. Bus. **74**(1), 101–124 (2001)

39. Stepanova, M., Thomas, L.: Survival analysis methods for personal loan data. Oper. Res. **50**(2), 277–289 (2002)
40. Sun, K., Huang, S.H., Wong, D.S.H., Jang, S.S.: Design and application of a variable selection method for multilayer perceptron neural network with lasso. IEEE Trans. Neural Netw. Learn. Syst. **28**(6), 1386–1396 (2016)
41. Tang, Z., Shen, Y., Zhang, X., Yi, N.: The spike-and-slab lasso Cox model for survival prediction and associated genes detection. Bioinformatics **33**(18), 2799–2807 (2017)
42. Tibshirani, R.: Regression shrinkage and selection via the lasso. J. R. Stat. Soc. Ser. B (Methodol.) **58**(1), 267–288 (1996)
43. Tibshirani, R.: The lasso method for variable selection in the Cox model. Stat. Med. **16**(4), 385–395 (1997)
44. Tibshirani, R.: Regression shrinkage and selection via the lasso: a retrospective. J. R. Stat. Soc. Ser. B (Stat. Methodol.) **73**(3), 273–282 (2011)
45. Zellner, D., Keller, F., Zellner, G.E.: Variable selection in logistic regression models. Commun. Stat. Simul. Comput. **33**(3), 787–805 (2004)

Operations with Iso-structured Models with Commutative Orthogonal Block Structure: An Introductory Approach

Carla Santos, Cristina Dias, Célia Nunes, and João T. Mexia

Abstract An approach to models based on an algebraic context allows interesting and useful statistical results to be derived or at least better understood. In the approach to models with commutative orthogonal block structure via algebraic structure it is possible to show that the orthogonal projection matrix in the space spanned by the mean vector commuting with the covariance matrix guarantees least squares estimators giving best linear unbiased estimators for estimable vectors. In this work we focus on the possibility of performing operations with models with commutative orthogonal block structure that are iso-structured, that is, models generating the same commutative Jordan Algebra of symmetric matrices.

Keywords Algebraic structure · Cartesian product · Jordan Algebra · Linear mixed model

C. Santos (✉)
Polytechnic Institute of Beja, Campus IPBeja, Beja, Portugal

NOVAMath-Center for Mathematics and Applications, SST, New University of Lisbon, Caparica Campus, Caparica, Portugal
e-mail: carla.santos@ipbeja.pt

C. Dias
Polytechnic Institute of Portalegre, Campus Politécnico, Portalegre, Portugal

NOVAMath-Center for Mathematics and Applications, SST, New University of Lisbon, Caparica Campus, Caparica, Portugal
e-mail: cpsd@ipportalegre.pt

C. Nunes
Department of Mathematics and Center of Mathematics and Applications, University of Beira Interior, Covilhã, Portugal
e-mail: celian@ubi.pt

J. T. Mexia
NOVAMath-Center for Mathematics and Applications, SST, New University of Lisbon, Caparica Campus, Caparica, Portugal
e-mail: jtm@fct.unl.pt

© The Author(s), under exclusive license to Springer Nature Switzerland AG 2023

C. P. Kitsos et al. (eds.), *Statistical Modelling and Risk Analysis*,
Springer Proceedings in Mathematics & Statistics 430,
https://doi.org/10.1007/978-3-031-39864-3_13

1 Introduction

Linear mixed models stand out among the statistical tools for their versatility and power in the analysis of experimental data in several fields, such as agricultural research, medical research, and others, due to their suitability for correlated data.

A particular class of linear mixed models, named models with commutative orthogonal block structure (COBS) is interesting for the possibility of obtaining least squares estimators giving best linear unbiased estimators for estimable vectors.

As argued by Malley [12], the approach based on an algebraic context allows many interesting and useful statistical results to be derived or at least better understood. One possible such algebraic context involves Jordan Algebras (JA). In our algebraic approach to COBS, the central role is played by commutative Jordan Algebras of symmetric matrices (CJAS) since these algebras provide a refined discussion of the algebraic structure of the models.

The study of COBS through an approach based on its algebraic structure leads us to interesting results on the estimation of variance components and in the construction of models (see [9]) and facilitates the procedures associated with the operations with models that allow the joint study of models obtained separately.

The designs in an experimental network may be iso-structured to ensure robustness of the conclusions. Thus, if designs having been carried out on different "environments" the significant results obtained for them do not depend of the "environments". In this way we obtain results with a wide range of applicability. For instance, in plant breeding we aim at selecting cultivars with good performance in wide regions. This may be achieved with networks of designs for comparison of cultivars (see [10]).

In this work we focus on the possibility of performing operations between models when the initial models are iso-structured models, that is, models that correspond to experiments carried out with the same design.

The study of operations with models, using JA, had as its starting point the work of [8], which focused on the binary operations defined in the principal bases of the JA associated with the models. In the works by Mexia et al. [14] and Santos et al. [23], the possibility of joint analysis of several models is addressed, through operations between models based on their algebraic structure.

For Model Crossing and Model Nesting, studied in [14], COBS and fixed effects models were considered. In [20] the study of the Model Nesting operation involved mixed models and fixed effects models.

Another operation with models, called Model Joining, was introduced by Santos et al. [23] involving COBS, based on the Cartesian product of CJAS.

This paper is structured as follows. Since our approach to COBS lies in their algebraic structure resting in commutative Jordan Algebras of symmetric matrices, in Sect. 2 we provide some results on these structures. In Sect. 3 we present the succession of conditions that leads to the special class of mixed models constituted by the COBS. Section 4 is devoted to the operations with COBS, and to the possibility of performing these operations with COBS that generate the same CJAS. In Sect. 5 we present some concluding remarks.

2 Commutative Jordan Algebras of Symmetric Matrices

The structures known today as Jordan Algebras, originally called "r number systems", were introduced by Pascual Jordan to formalize the notion of an algebra of observables in Quantum Mechanics and developed in partnership with John von Neumann and Eugene Wigner, see [11]. Later on, these structures were rediscovered by Seely [25], who called them quadratic vector spaces and used them to solve statistical inference problems. With Seely was initiated a very fruitful research line with relevant developments of linear statistical inference, see [9, 15, 16, 21, 26–30, 32]. Among these, we would like to highlight the contribution of [15] and [16], who used Jordan Algebras in hypothesis testing, first for variance components and later for linear combinations of parameters in mixed linear models.

Following a path that will lead us to one of the focal points of our work, the commutative Jordan Algebras of symmetric matrices, let us start by defining an algebra, A, as a linear space provided with a binary operation, here denoted by $*$, usually called product, that satisfies the conditions,

$$\begin{cases} \alpha\,(x * y) = (\alpha x) * \ y = x * (\alpha y) \\ (x + y) * z = x * z + y * z \\ x * (y + z) = x * \ y + x * z \\ \alpha\,(x + y) = \alpha x + \alpha y \end{cases}$$

for all $\alpha \in \mathbb{R}, x, y, z \in A$ [12].

The product $*$ enjoys the associative and commutative properties, however these properties are not necessary for a linear space to be an algebra.

An algebra A is said to be an associative algebra when

$$x * \ (y * z) = (x * \ y) * z$$

and a commutative algebra when

$$x * y = y * x$$

for all $x, y, z \in A$.

A Jordan Algebra (JA) is a commutative algebra whose product satisfies the Jordan identity, given by

$$x^2 * \ (y * z) = \left(x^2 * y\right) * \ z,$$

with $x^2 = x * x$, for all $x, y, z \in A$.

Note that a JA does not have to be an associative algebra but must obey the Jordan identity, which constitutes a more restricted type of associativity.

When the matrices of a JA commute it is called a commutative Jordan Algebra (CJA).

Since there are linear spaces constituted by matrices, closed for the Jordan product of matrices, and containing the squares of their matrices that, even if their matrices commute, are isomorphic to no CJA constituted by symmetric matrices [12], we will consider only CJA constituted by symmetric matrices (CJAS).

To summarize what was previously set, we can say that a CJAS is a linear space constituted by symmetric matrices that commute containing the squares of their matrices. As shown by Seely [27], every CJAS, A, has a unique basis, the principal basis, $pb(A)$, constituted by pairwise orthogonal orthogonal projection matrices (POOPM).

Let $pb\,(A) = \{\mathbf{Q}_1, \ldots, \mathbf{Q}_m\}$. Given $\mathbf{M}$ a matrix belonging to A, we have

$$\mathbf{M} = \sum_{j=1}^{m} b_j \mathbf{Q}_j = \sum_{j \in C(\mathbf{M})} b_j \mathbf{Q}_j$$

with $C\,(\mathbf{M}) = \left\{j : b_j \neq 0\right\}$.

Since the Moore-Penrose inverse of $\mathbf{M}$ is

$$\mathbf{M}^+ = \sum_{j=1}^{m} b_j^+ \mathbf{Q}_j$$

where $b_j^+ = b_j^{-1}$, for all $b_j \neq 0$, $j = 1, \ldots, m$, and so

$$C\left(\mathbf{M}^+\right) = C\,(\mathbf{M}),$$

a CJAS contains the Moore-Penrose inverses of any of its matrices.

With $\nabla_j = R\left(\mathbf{Q}_j\right)$, $j = 1, \ldots, m$, and $g_j = \text{rank}\left(\mathbf{Q}_j\right)$, $j = 1, \ldots, m$, representing by $\oplus$ the orthogonal direct sum of subspaces, we have

$$\begin{cases} R\,(\mathbf{M}) = \bigoplus_{j \in C(\mathbf{M})} \nabla_j \\ \\ r\,(\mathbf{M}) = rank\,(\mathbf{M}) = \sum_{j \in C(\mathbf{M})} g_j \end{cases}$$

Moreover, the orthogonal projection matrix on $R\,(\mathbf{M})$ will be

$$\mathbf{Q}\,(\mathbf{M}) = \sum_{j \in C(\mathbf{M})} \mathbf{Q}_j.$$

Given $\mathbf{Q}$, an orthogonal projection matrix belonging to A, we have

$$\mathbf{Q} = \sum_{j=1}^{m} b_j \mathbf{Q}_j.$$

Since $\mathbf{Q}$ is idempotent and $\mathbf{Q}_1, \ldots, \mathbf{Q}_m$ are idempotent and pairwise orthogonal,

$$\mathbf{Q} = \sum_{j=1}^{m} b_j \mathbf{Q}_j = \sum_{j=1}^{m} b_j^2 \mathbf{Q}_j = \mathbf{Q}^2,$$

coming $b_j^2 = b_j$ and so $b_j = 0$ or $b_j = 1$, $j = 1, \ldots, m$, then the orthogonal projection matrices belonging to a CJAS, A, are sums of matrices of the $pb\,(A)$, that is, with $C\,(\mathbf{Q}) = \left\{j : b_j \neq 0\right\}$,

$$\mathbf{Q} = \sum_{j \in C(\mathbf{Q})} \mathbf{Q}_j.$$

Since $pb\,(A) = \{\mathbf{Q}_1 \ldots, \mathbf{Q}_m\}$ has m matrices, A, as a linear subspace, has dimension $dim(A) = m$. Thus, since there are as many orthogonal projection matrices (OPM) in A as there are distinct sums of matrices of $pb\,(A)$, that can be 2^m OPM in A, once each of the sums corresponds to a sub-set of $\bar{\bar{m}} = \{1, \ldots, m\}$. Given $C \subseteq \bar{\bar{m}}$,

$$\mathbf{Q}\,(C) = \sum_{j \in C} \mathbf{Q}_j$$

so that, with $r\,(C) = rank\,(\mathbf{Q}\,(C))$, we will have $r\,(C) = \sum_{j \in C} g_j$.

Given the family $M = \{\mathbf{M}_1, \ldots, \mathbf{M}_w\}$ of matrices of A, we will have

$$\mathbf{M}_i = \sum_{j=1}^{m} b_{i,j} \mathbf{Q}_j, \; i = 1, \ldots, w$$

and $B = \left[b_{i,j}\right]$ will be the transition matrix between M and Q, $M \backslash Q$. The matrices in M are linearly independent when and only when the row vectors of B are linearly independent.

Since $dim(A) = m$, if $w = m$ and the matrices $\mathbf{M}_1, \ldots, \mathbf{M}_m$ are linearly independent the m row vectors of B will be linearly independent, thus B will be $m \times m$ and rank $(B) = m$. Then B will be invertible and with $B^{-1} = \left[b^{l,h}\right]$ we will have

$$\mathbf{Q}_l = \sum_{h=1}^{m} b^{l.h} \mathbf{M}_m, \; l = 1, \ldots, m$$

and $M = \{\mathbf{M}_1, \ldots, \mathbf{M}_w\}$ will be a basis for A.

Now, the matrices of the family $M = \{\mathbf{M}_1, \ldots, \mathbf{M}_w\}$ commute if and only if they are diagonalized by the same matrix, $\mathbf{P}^o$. We then have $M \subset V(\mathbf{P}^o)$, with $V(\mathbf{P}^o)$ the family of matrices diagonalized by $\mathbf{P}^o$. Since $V(\mathbf{P}^o)$ is a CJAS, we see that a family of $n \times n$ symmetric matrices is contained in a CJAS if and only if they commute. Since the intersection of CJAS gives CJAS there will be a minimum CJAS containing M, whose matrices commute, this will be the CJAS generated by M.

3 Models with Commutative Orthogonal Block Structure

Let us consider a mixed model given by

$$\mathbf{Y} = \sum_{i=0}^{w} \mathbf{X}_i \boldsymbol{\beta}_i$$

where $\boldsymbol{\beta}_0$ is fixed and $\boldsymbol{\beta}_1, \ldots, \boldsymbol{\beta}_w$ are independent random vectors with null mean vectors, covariance matrices $\sigma_1^2 \mathbf{I}_{c_1} \ldots \sigma_w^2 \mathbf{I}_{c_w}$, where $c_i = rank\,(\mathbf{X}_i)\,, i = 1, \ldots, w$. The matrices $\mathbf{X}_1, \cdots, \mathbf{X}_w$ are known and such that $R\,([\mathbf{X}_1 \cdots \mathbf{X}_w]) = \mathbb{R}^n$.

The model $\mathbf{Y}$ has mean vector

$$\boldsymbol{\mu} = \mathbf{X}_0 \boldsymbol{\beta}_0$$

and covariance matrix

$$\mathbf{V} = \sum_{i=1}^{w} \sigma_i^2 \mathbf{M}_i,$$

where $\mathbf{M}_i = \mathbf{X}_i \mathbf{X}_i^T, \ i = 1, \ldots, w$.

The space, Ω, spanned by $\boldsymbol{\mu}$ will be $R(X_0)$, so the orthogonal projection matrix on Ω will be

$$\mathbf{T} = \mathbf{X}_0 \left(\mathbf{X}_0^T \mathbf{X}_0\right)^+ \mathbf{X}_0^T = \mathbf{X}_0 \mathbf{X}_0^+.$$

When the matrices of the family $M = \{\mathbf{M}_1, \ldots, \mathbf{M}_w\}$ commute, they generate a CJAS, A. We say that this CJAS is generated by the COBS M, and we put $A = A(M)$.

The principal basis of the CJAS A, $pb\,(A)$, is constituted by pairwise orthogonal orthogonal projection matrices (POOPM), $Q_i, i = 1, \ldots, m$ [28].

Putting

$$\mathbf{M}_i = \sum_{j=1}^{m} b_{i,j} \mathbf{Q}_j, \ i = 1, \ldots, w,$$

the variance-covariance matrix, $\mathbf{V}$, can be written as the linear combination of the matrices of the $pb\,(A)$,

$$\mathbf{V} = \sum_{j=1}^{m} \gamma_j \mathbf{Q}_j,$$

with $\gamma_j = \sum_{i=1}^{w} b_{i,j}\sigma_i^2,\ j = 1, \ldots, m^0$, the canonical variance components.

In the framework of the design of experiments in agricultural trials, [17, 18] introduced a new class of mixed models for which the matrices of the $pb\,(A)$ add up to the identity matrix,

$$\sum_{j=1}^{m} \mathbf{Q}_j = \boldsymbol{I}_n.$$

This class of mixed models, named models with orthogonal block structure (OBS), took a central part in the theory of randomized block designs, see e.g. [1, 2]. OBS allow optimal estimation for variance components of blocks and contrasts of treatments, however, inference in OBS usually involves orthogonal projections on the range spaces of the matrices $\mathbf{Q}_j,\ j = 1, \ldots, m$, which comprises some complexity due to the combination of estimators obtained from different projections. A more restricted class of mixed models introduced by Fonseca et al. [9], named models with commutative orthogonal block structure (COBS), allows to overcome this obstacle, and achieve least squares estimators (LSE) giving best linear unbiased estimators (BLUE) for estimable vectors.

A mixed model is COBS if it is OBS and, moreover, $\mathbf{T}$, the orthogonal projection matrix on the space spanned by the mean vector, commutes with the matrices $\mathbf{Q}_1, \ldots, \mathbf{Q}_m$ (see e.g. [9])

$$\mathbf{T}\mathbf{Q}_j = \mathbf{Q}_j\mathbf{T},\ j = 1, \ldots, m.$$

Noting that the matrices $\mathbf{T}$ and $\mathbf{Q}_1, \ldots, \mathbf{Q}_m$ will belong to the CJAS, A, generated by the matrices $\mathbf{M}_i,\ i = 1, \ldots, w$, the model is COBS when the matrices $\mathbf{M}_1, \ldots, \mathbf{M}_w$ and $\mathbf{T}$ commute.

As showed by Zmyślony [31], the commutativity between $\mathbf{T}$ and the covariance matrix,$\mathbf{V}$, is a necessary and sufficient condition for least square estimators (LSE) to be best linear unbiased estimators (BLUE). A general condition for the commutativity between $\mathbf{T}$ and $\mathbf{V}$, resorting to U-matrices, was introduced by Santos et al. [24], using the fundamental partition of $\mathbf{Y}$, constituted by the sub-vectors $\mathbf{Y}_1, \ldots, \mathbf{Y}_{\dot{n}}$, corresponding to the $\dot{n}$ sets of the levels of the fixed effects factors.

As pointed out by Fonseca et al. [9] the approach to COBS based on its algebraic structure leads to interesting results on the estimation of variance components and on the building up of models, so this approach has been explored in several works, which, among other aspects, have focused on inference (e.g. [3, 4, 6, 7, 13, 19]), the relationship between COBS and other models (e.g. [5, 22]) and operations with models (e.g. [14, 23]).

4 Operating Iso-structured COBS

In experimental designs the two most common building blocks are crossed factors and nested factors. We say that factors are crossed when every level of one factor occurs with every level of the other factors, and that a factor is nested within another factor when any given level of the nested factor appears at only one level of the nesting factor, that is, when the levels of the nested factor are divided among the levels of the nesting factor.

Taking as a starting point crossed factors and nested factors, the operations of Model Crossing and Model Nesting were defined (see [14]). These operations with models are based on binary operations on the principal bases of the Jordan algebras associated with the models (see [8]).

Equivalent to crossing and nesting factors in a model with u factors, we can consider u models and perform crossing or nesting with those models. Considering each one of the u models with only one factor with $a_1, \ldots, a_u$ levels, when we cross these models, we obtain the same combination of levels we would have in a single model with u crossed factors, with $a_1, \ldots, a_u$ levels, thus the same number of treatments. In a similar approach to that described for the Model Crossing, performing the operation of Model Nesting involving u models, each one with only one factor, is equivalent to nesting of u factors of a single model. In a generalization of these operations, we can perform the operations of Model Crossing and Model Nesting with several models, each one of them with more than one factor [24].

Focusing on the objective of this work, that is to establish a preamble to operations with models from a family of models that correspond to experiments performed with the same design, we highlight that the use of such models (with the same algebraic structure and independent observation vectors), in addition to allowing the systematization of the inference, facilitates the comparative study of the results of the various experiments and their possible integration in synthesis works.

Naming models with the same algebraic structure and independent observation vectors as iso-structured (ISO), we say that two COBS $\mathcal{M}_1$ and $\mathcal{M}_2$ are ISO if they generate the same CJAS. Thus, an equivalence relationship is defined in the space of the COBS, and we put

$$\mathcal{M}_1 \ \tau \ \mathcal{M}_2.$$

Let us now consider two pairs of COBS $(\mathcal{M}_{l,1}, \mathcal{M}_{l,2}), l = 1, 2,$ such that

$$\mathcal{M}_{1,h} \ \tau \ \mathcal{M}_{2,h}, h = 1, 2,$$

having, therefore,

$$A\left(\mathcal{M}_{1,h}\right) = A\left(\mathcal{M}_{2,h}\right)$$

as well as

$$A\left(\mathcal{M}_{1,1}\right) \bigotimes A\left(\mathcal{M}_{1,2}\right) = A\left(\mathcal{M}_{2,1}\right) \bigotimes A\left(\mathcal{M}_{2,2}\right)$$

and consequently

$$\mathcal{M}_{1,1} \bigotimes \mathcal{M}_{1,2} \ \ \tau \ \ \mathcal{M}_{2,1} \bigotimes \mathcal{M}_{2,2},$$

what shows that the relation ISO is a congruence relation for the models product. This result extends directly to the Cartesian product. Let

$$pb\left(A_l\right) = \left\{\mathbf{Q}_{l,1}, \ldots, \mathbf{Q}_{l,m_l}\right\}$$

be the principal basis of de CJAS A_l. Then the principal basis of $\underset{l=1}{\overset{u}{\times}} A_l$ is formed by the block diagonal matrix in which the principal sub-matrices are $\mathbf{0}_{n_l \times v_l}$, except for the h-th that belongs to the principal basis $pb\left(A_l\right), l = 1, \ \ldots, \ u$. Let us now consider that we have

$$\mathcal{M}_{1,v} \ \ \tau \ \ \mathcal{M}_{2,v}, v = 1, \ \ldots, \ u$$

so that

$$A\left(\mathcal{M}_{1,v}\right) = A\left(\mathcal{M}_{2,v}\right), v = 1, \ \ldots, \ u$$

and consequently

$$\underset{v=1}{\overset{u}{\times}} \ A\left(\mathcal{M}_{1,v}\right) = \underset{v=1}{\overset{u}{\times}} \ A\left(\mathcal{M}_{2,v}\right)$$

thus

$$\overset{u}{\underset{v=1}{\times}} \mathcal{M}_{1,v} \;\tau\; \overset{u}{\underset{v=1}{\times}} \mathcal{M}_{2,v}.$$

So, the relation ISO continues to behave as a congruence for the Cartesian product of models. We say that a COBS, $\mathcal{M}$, is regular and complete if it generates a complete and regular CJA. Thus, if the matrices of the model $\mathcal{M}$ are $n \times n$, with $pb\left(A\left(\mathcal{M}\right)\right) = \{\mathbf{Q}_1, \ldots, \mathbf{Q}_m\}$ we will have

$$\begin{cases} \mathbf{Q}_1 = \frac{1}{n}\,\mathbf{1}_n\,\mathbf{1}_n^T \\ \sum_{j=1}^{m} \mathbf{Q}_j = \mathbf{I}_n \end{cases}.$$

Given the regular and complete models $\mathcal{M}_1$ and $\mathcal{M}_2$, with matrices $n_1 \times n_1$ and $n_2 \times n_2$, if we nest the second model in the first, we get the model

$$\mathcal{M}_1 * \mathcal{M}_2$$

with

$$pb\left(A\left(\mathcal{M}_1 * \mathcal{M}_2\right)\right) =$$

$$= \left\{\mathbf{Q}_{1,1} \otimes \frac{1}{n_2}\,\mathbf{1}_{n_2}\,\mathbf{1}_{n_2}^T, \;\ldots,\; \mathbf{Q}_{1,m_1} \otimes \frac{1}{n_2}\,\mathbf{1}_{n_2}\,\mathbf{1}_{n_2}^T,\; \mathbf{I}_{n_1} \otimes \mathbf{Q}_{2,2},\; \ldots,\; \mathbf{I}_{n_1} \otimes \mathbf{Q}_{2,m_2}\right\}$$

where $\otimes$ denotes the Kronecker matrix product, and

$$pb\left(A\left(\mathcal{M}_v\right)\right) = \left\{\mathbf{Q}_{v,1}, \ldots, \mathbf{Q}_{v,m_v}\right\},\; v = 1, 2.$$

Note that when $\mathcal{M}_1$ and $\mathcal{M}_2$ are regular and complete models, $\mathcal{M}_1 * \mathcal{M}_2$ will be regular and complete. If we restrict ourselves now to regular and complete COBS, and consider $\mathcal{M}_{1,v}\ \tau\ \mathcal{M}_{2,v}$, $v = 1, 2$, it turns out that we will have

$$\left(\mathcal{M}_{1,1} * \mathcal{M}_{1,v}\right)\ \tau\ \left(\mathcal{M}_{2,1} * \mathcal{M}_{2,v}\right)$$

and $*$ behaving as a congruence.

5 Conclusion

In this work we focused on the possibility of performing operations involving models with commutative orthogonal block structure when these models are ISO, that is, when they generate the same commutative Jordan Algebra of symmetric matrices. We have showed that the relation ISO behaves as a congruence for products of models.

Acknowledgments This work was partially supported by the Fundação para a Ciência e a Tecnologia (Portuguese Foundation for Science and Technology) through the projects UIDB/00297/2020 and UIDB/00212/2020.

References

1. Caliński, T., Kageyama, S.: Block Designs: A Randomization Approach. Vol. I: Analysis, Lecture Note in Statistics, vol. 150. Springer, New York (2000)
2. Caliński, T., Kageyama, S.: Block Designs: A Randomization Approach. Vol. II: Design, Lecture Note in Statistics, vol. 170. Springer, New York (2003)
3. Carvalho, F., Mexia, J.T, Oliveira, M.: Canonic inference and commutative orthogonal block structure. Discuss. Math. Probab. Stat. **28**(2), 171–181 (2008)
4. Carvalho, F., Mexia, J.T, Oliveira, M.: Estimation in models with commutative orthogonal block structure. J. Stat. Theory Practice **3**(2), 525–535 (2009). https://doi.org/10.1080/15598608.2009.10411942
5. Carvalho, F., Mexia, J.T, Santos, C.: Commutative orthogonal block structure and error orthogonal models. Electron. J. Linear Algebra **25**, 119–128 (2013). https://doi.org/10.13001/1081-3810.1601
6. Carvalho, F., Mexia, J.T, Santos, C., Nunes, C.: Inference for types and structured families of commutative orthogonal block structures. Metrika **78**, 337–372 (2015). https://doi.org/10.1007/s00184-014-0506-8
7. Ferreira, S., Ferreira, D., Nunes, C., Carvalho, F., Mexia, J.T.: Orthogonal block structure and uniformly best linear unbiased estimators. In: Ahmed, S., Carvalho, F., Puntanen, S. (eds.) Matrices, Statistics and Big Data. IWMS 2016. Contributions to Statistics. Springer (2019)
8. Fonseca, M., Mexia, J. T., Zmyślony,R.: Binary operations on Jordan algebras and orthogonal normal models. Linear Algebra Appl. **117**(1), 75–86 (2006). https://doi.org/10.1016/j.laa.2006.03.045
9. Fonseca, M., Mexia, J.T., Zmyślony, R.: Inference in normal models with commutative orthogonal block structure. Acta Comment. Univ. Tartu. Math. **12**, 3–16 (2008)
10. Gusmão, L., Mexia, J.T., d Gomes, M.L.: Mapping of equipotential zones for cultivar yield pattern evolution. Plant Breed. **103**, 293–298 (1989)
11. Jordan, P., Von Neumann, J., Wigner, E.: On an algebraic generalization of the quantum mechanical formulation. Ann. Math. **35**(1), 29–64 (1934). https://doi.org/10.2307/1968117
12. Malley, J.D.: Statistical Applications of Jordan Algebras. Lecture Notes in Statistics, vol. 91. Springer, New York (1994)
13. Mexia, J.T., Nunes, C., Santos, C.: Structured families of normal models with COBS. In: 17th International Workshop in Matrices and Statistics, 23–26 July, Tomar (Portugal), Conference paper (2008)
14. Mexia, J.T., Vaquinhas, R., Fonseca, M., Zmyślony, R.: COBS: segregation, matching, crossing and nesting. In: Latest Trends and Applied Mathematics, Simulation, Modelling, 4th

International Conference on Applied Mathematics, Simulation, Modelling, ASM'10, pp. 249–255 (2010)
15. Michalski, A., Zmyślony, R.: Testing hypothesis for variance components in mixed linear models. Statistics **27**, 297–310 (1996). https://doi.org/10.1080/02331889708802533
16. Michalski, A., Zmyślony, R.: Testing hypothesis for linear functions of parameters in mixed linear models. Tatra Mountain Math. Publ. **17**, 103–110 (1999)
17. Nelder, J.A.: The analysis of randomized experiments with orthogonal block structure I, Block structure and the null analysis of variance. Proc. R. Soc. Lond. Ser. A Math. Phys. Eng. Sci. **283**, 147–162 (1965)
18. Nelder, J.A.: The analysis of randomized experiments with orthogonal block structure II. Treatment structure and the general analysis of variance. Proc. R. Soc. Lond. Ser. A Math. Phys. Eng. Sci. **283**, 163–178 (1965)
19. Nunes, C., Santos, C., Mexia, J.T.: Relevant statistics for models with commutative orthogonal block structure and unbiased estimator for variance components. J. Interdiscip. Math. **11**, 553–564 (2008). https://doi.org/10.1080/09720502.2008.10700581
20. Ramos, P., Fernandes, C., Mexia, J.T.: Algebraic structure for interaction on mixed models. J. Interdiscip. Math. **18**(1–2), 43–52. https://doi.org/10.1080/09720510.2014.927622
21. Rao, C., Rao, M.: Matrix Algebras and Its Applications to Statistics and Econometrics. World Scientific (1998)
22. Santos, C., Nunes, C., Mexia, J.T.: OBS, COBS and mixed models associated to commutative Jordan Algebra. Bulletin of the ISI, LXII, Proceedings of 56th session of the International Statistical Institute, Lisbon, pp. 3271–3274 (2008)
23. Santos, C., Nunes, C., Dias, C., Mexia, J.T.: Joining models with commutative orthogonal block structure. Linear Algebra Its Appl. **517**, 235–245 (2017). https://doi.org/10.1016/j.laa.2016.12.019
24. Santos, C., Dias, C., Nunes, C., Mexia, J.T.: On the derivation of complex linear models from simpler ones. In: Proceedings of the 5th NA International Conference on Industrial Engineering and Operations Management Detroit, Michigan, USA, August 10–14, 2020, pp. 603–613 (2020)
25. Seely, J.: Linear spaces and unbiased estimation. Ann. Math. Stat. **41**(5), 1725–1734 (1970). https://doi.org/10.1214/aoms/1177696818
26. Seely, J.: Quadratic subspaces and completeness. Ann. Math. Stat. **42**(2), 710–721 (1971). https://doi.org/10.1214/aoms/1177693420
27. Seely, J.: Minimal sufficient statistics and completeness for multivariate normal families. Sankhya **39**, 170–185 (1977)
28. Seely, J., Zyskind, G.: Linear Spaces and minimum variance estimation. Ann. Math. Stat. **42**(2), 691–703 (1971). https://doi.org/10.1214/aoms/1177693418
29. Vanleeuwen, D., Seely, J., Birkes, D.: Sufficient conditions for orthogonal designs in mixed linear models. J. Stat. Plan. Inference **73**, 373–389 (1998). https://doi.org/10.1016/S0378-3758(98)00071-8
30. Vanleeuwen, D., Birkes, D., Seely, J.: Balance and orthogonality in designs for mixed classification models. Ann. Stat. **27**(6), 1927–1947 (1999). https://doi.org/10.1214/aos/1017939245
31. Zmyślony, R.: A characterization of best linear unbiased estimators in the general line-ar model. In: Lecture Notes in Statistics, vol. 2, pp. 365–373 (1978)
32. Zymślony, R., Drygas, H.: Jordan algebras and Bayesian quadratic estimation of variance components. Linear Algebra Appl. **168**, 259–275 (1992). https://doi.org/10.1016/0024-3795(92)90297-N

Long and Short–Run Dynamics in Realized Covariance Matrices: A Robust MIDAS Approach

Scaffidi Domianello Luca and Edoardo Otranto

Abstract A recent stream of the econometric literature is devoted to modelize unobservable short and long–run components in volatility and time–varying correlations of financial assets. In such models two typical problems are the sensitivity of the estimation results to the order in which the assets enter the model and the trade-off between the flexibility of the model and its parsimony. We propose a new class of additive component models belonging to the MIDAS family, that overcomes some drawbacks related to the use of the Cholesky decomposition of the covariance matrix, avoiding the effect of the asset order on the estimation process. Moreover, we deal with the *curse of dimensionality problem* by adopting the Hadamard exponential function which allows asset-pair-specific and time-varying parameters. We verify the advantage of the proposed models by comparing them with some benchmarks, in terms of both in–sample and out–of–sample performance, through some statistical and economic loss functions.

Keywords Realized covariance · MIDAS · Dynamic component models · Hadamard exponential parameterization

1 Introduction

Forecasting time–varying conditional (co)variances is a widely studied topic in financial literature, due to the importance of volatility of asset returns and their correlation for financial applications, such as: hedging, asset allocation, pricing, risk management, etc.

L. Scaffidi Domianello (✉)
University of Florence, Department of Statistics, Computer Science, Applications, Florence, Italy
e-mail: luca.scaffididomianello@unifi.it

E. Otranto
University of Messina, Department of Economics, Messina, Italy
e-mail: eotranto@unime.it

© The Author(s), under exclusive license to Springer Nature Switzerland AG 2023
C. P. Kitsos et al. (eds.), *Statistical Modelling and Risk Analysis*,
Springer Proceedings in Mathematics & Statistics 430,
https://doi.org/10.1007/978-3-031-39864-3_14

Early multivariate volatility models (e.g. the BEKK of [14]) were based on daily cross–product returns and assumed a constant average (or long–run) level of (co)variances, though empirical evidence suggests that it is time–varying (see, for example, the results in [15], for the S&P 500 volatility index).

In the last decade, a great deal of effort was put into the development of models based on Realized Covariance matrix (see, for example, the Conditional Autoregressive Wishart—CAW—model of [17]), modeling directly a nonparametric estimation of the Covariance matrices, based on intra–daily returns.

A relatively recent stream of literature is related to long and short–run components that characterize, with different dynamics, the Realized Covariance series (see, for example, [7]). By decomposing the Conditional Covariance matrix into a short–run and a long–run component, it is possible to capture, in a parsimonious way, the long–memory behavior of (co)variances. The short–run component is aimed to capture daily fluctuations and transitory effects; conversely, the long–run component represents the average level that varies over time according to economic conditions. However, dynamic component models are based on the Cholesky decomposition, which makes the short–run component potentially sensible to asset order. Furthermore, models require suitable parameterizations to guarantee the positive definiteness of the estimated covariance matrices and a small number of unknown coefficients to avoid the so–called *curse of dimensionality problem*.

We propose a new class of additive component models, belonging to the MIxed DAta Sampling, or MIDAS [16], family in the spirit of [10], with features that help us overcome some drawbacks:

- it does not depend on the Cholesky decomposition of the Covariance matrix, so that the order of the series is not relevant in the estimation of the model parameters;
- the multiplicative decomposition of the covariance matrix, adopted in other models, requires the calculation at each time of the inverse of the Cholesky factor, thus slowing down the optimization algorithm. Our additive specification does not require this step, with a clear computational gain;
- multivariate volatility models, to overcome the *curse of dimensionality problem*, usually assume a scalar specification of the conditional (co)variances, imposing the same dynamics for each series. This hypothesis is very strong and not supported by empirical evidence. The model we introduce adopts the Hadamard exponential function proposed by [5], which allows asset–pair specific and time–varying parameters. This specification offers a more flexible dynamics with only one more parameter than the baseline specification, thus preserving the parsimony of the model.

In next section we introduce the models we propose, highlighting their robustness to the order of assets and its flexibility; in the same section we provide the lines for parameter estimation. Section 3 contains the empirical analysis, where we fit a set of models (the proposed models and the benchmarks) to the Realized Covariance series of 9 assets belonging to the Dow Jones Industrial Average (DJIA) index; we compare the in–sample fitting and the out–of–sample performance of the estimated models in terms of statistical and economic loss functions. Some final remarks will conclude the paper.

2 A New MIDAS-Type Model

Let C_t be an n–order Realized Covariance matrix, that is assumed to follow an n-dimensional conditional Wishart distribution:

$$C_t|\mathcal{I}_{t-1} \sim W_n\left(\nu, S_t/\nu\right), \quad \forall\, t = 1, \ldots, T \tag{1}$$

where $\mathcal{I}_{t-1}$ is the information set at time $t-1$, $\nu > n-1$ are the degrees of freedom, while S_t is a positive definite simmetric (PDS) scale matrix and it is the conditional expectation of the Realized Covariance matrix, (C_t):

$$E(C_t|\mathcal{I}_{t-1}) = S_t \tag{2}$$

Let us consider a scalar BEKK-type dynamics for the Conditional Covariance matrix, S_t:

$$S_t = M(1-\alpha-\beta) + \alpha C_{t-1} + \beta S_{t-1} \tag{3}$$

where M is the PDS unconditional covariance matrix; we impose the following sufficient constraints to ensure the stationarity of the process: $\alpha \geq 0$, $\beta \geq 0$, $\alpha + \beta < 1$. Similarly to [10], we allow the intercept matrix, M, to be time–varying.[1] Then, we additively decompose the Conditional Covariance matrix, S_t, into a slow–moving long–run component and a short–run component. The Realized BEKK MIDAS (ReBEKKMIDAS) is specified as follows:

$$S_t = M_t(1-\alpha-\beta) + \alpha C_{t-1} + \beta S_{t-1} \tag{4}$$

[1] [10] assumed a DCC-type dynamics for the Quasi Conditional Correlation equation, by replacing the constant intercept matrix with a time–varying one.

The long–run component, M_t, representing the time–varying average level of the Conditional Covariance, is specified as follows:

$$M_t = \bar{\Lambda} + \theta \sum_{k=1}^{K} \varphi_k(\omega_1, \omega_2) C_{t-k}^{(m)}$$
$$C_{t-k}^{(m)} = \sum_{\tau=t-mk}^{t-m(k-1)-1} C_\tau \tag{5}$$
$$\varphi_k(\omega_1, \omega_2) = \frac{(k/K)^{\omega_1-1}(1-k/K)^{\omega_2-1}}{\sum_{j=1}^{K}(j/K)^{\omega_1-1}(1-j/K)^{\omega_2-1}}$$

where, $\bar{\Lambda} = LL'$ with L an n–order lower triangular matrix, with positive diagonal entries as identifying condition; $C_t^{(m)}$ is an n–order matrix of monthly Realized Covariances, i.e., the aggregation of daily Realized Covariances over a period of 22 days.[2] $\varphi(\omega_1, \omega_2)$ is a weighting function of the past K values of $C_t^{(m)}$, with weights summing to one; if, in general, $\omega_1 = 1$ and $\omega_2 > 1$, this function is monotonically decreasing. θ is a non-negative scalar parameter aimed at capturing the effect of the weighted sum of the K past monthly Realized Covariance matrices on the long–run component.

For what concerns the latter, in Eq. (4) we have a unique weighting scheme and slope coefficient for variances and covariances, while in [10] there is a different specification for the variances and correlations. Furthermore, empirical evidence suggests that the pattern of the long–run component is very similar among the assets (see [2]); for example, periods of high volatility, leading to an increase in the average level, are almost equal for all the series. By using the same coefficients we avoid the proliferation of parameters on the estimation process.

The dynamics of S_t captures short–lived effects: indeed, if we rewrite Eq. (4) we obtain

$$S_t - M_t = \alpha(C_{t-1} - M_t) + \beta(S_{t-1} - M_t) \tag{6}$$

which shows as the short-run component fluctuates around the long-run one, M_t.

Furthermore, through the parameterization we propose, the estimation is invariant to the order of the assets. Indeed, the multiplicative decomposition of the covariance matrix, adopted in other models, such as the Multivariate MIDAS Aggregated Realized BEKK (MMAReBEKK) of [7],[3] makes the short-run Realized

[2] Notice that we allow the monthly Realized Covariance matrix to change day by day, while in principle it could be constant for the whole low–frequency period.

[3] Note that the MMAReBEKK is the scalar parameterization of the CAW MIDAS introduced by [17].

Covariance matrix potentially sensible to asset order.[4] Additionally, they require the calculation at each time of the inverse of the Cholesky factor of the long-run component, $M_t^{-1/2}$, thus slowing down the optimization algorithm. The additive specification we propose does not require this step, with a clear computational gain. In the empirical analysis in Sect. 3, we get a 30% reduction in estimation time.

2.1 The Hadamard Exponential Specification

The scalar ReBEKKMIDAS imposes the same dynamics for each asset, thus resulting in a very strong condition. For this purpose, we extend the model above through the Hadamard exponential function proposed by [4, 5], which allows asset–pair specific and time–varying coefficients, with one more parameter than the scalar parameterization.

Let us consider the scalar ReBEKK specified through the Hadamard parameterization:[5]

$$S_t = M(1-\alpha-\beta) + \alpha J_n \odot C_{t-1} + \beta J_n \odot S_{t-1} \tag{7}$$

where, J_n is an order n square matrix of ones, and $\odot$ is the element-by-element (Hadamard) product.

The Hadamard Exponential Realized BEKK MIDAS (HEReBEKK-MIDAS) model can be specified as follows:

$$S_t = M_t(1-\bar{\alpha}_t-\bar{\beta}_t) + A_t \odot C_{t-1} + B_t \odot S_{t-1} \tag{8}$$

where A_t and B_t are two matrices of the asset–pair specific and time–varying coefficients, parameterized through the Hadamard Exponential function; more in detail, A_t is parameterized as (similarly for B_t):

$$A_t = \alpha\, exp^{\odot}[\phi_A(N_{t-1} - J_n)] = \alpha \frac{exp^{\odot}[\phi_A(N_{t-1})]}{exp(\phi_A)} \tag{9}$$

where $exp^{\odot}$ is the Hadamard exponential function, or the entry–wise exponential operator,[6] ϕ_A is a scalar parameter, while N_t is equal to the Realized Correlation matrix $P_t = \{diag(C_t)\}^{-1/2}\, C_t\, \{diag(C_t)\}^{-1/2}$ or the Conditional Correlation matrix $R_t = \{diag(S_t)\}^{-1/2}\, S_t\, \{diag(S_t)\}^{-1/2}$.[7] [5] justify the dependence of the

---

[4] The short-run Realized Covariance matrix is the Realized Covariance matrix purged by the long-run component.

[5] Note that the specification in Eq. (7) is equivalent to that one in Eq. (3).

[6] if A is a square matrix of order n, then $exp^{\odot} A = (exp(a_{ij}))$.

[7] The choice among the two correlation matrices is purely an empirical question.

coefficients in the matrix A_t on the past Conditional (or Realized) Correlation matrix, through the volatility clustering phenomenon, a stylized fact that characterizes asset returns. Moreover, when there is a period of high market volatility, correlations and their persistence increase, but, in principle, in a different way for each pair of assets. The parameterization above captures this empirical regularity. Broadly speaking, when correlation increases, the impact of the lagged Realized Covariance on the future Conditional Covariance is stronger than for a lower level of correlation.

The diagonal elements of A_t are equal to α, while the off-diagonal elements are between 0 and α; in fact N_{t-1} is a matrix of ones along the main diagonal by construction, whereas its off-diagonal elements lie between -1 and 1. Notice that if $\phi_A = 0$, then $A_t = \alpha J_n$, i.e. the model reduces to the scalar ReBEKKMIDAS, obtaining a matrix with constant parameters. Importantly, the entry–wise exponential operator ensures the positive definiteness of any PD matrix [4].

In Eq. (8) an exact parameterization of the time-varying intercept term should be the following: $M_t \odot (J_n - A_t - B_t)$. Nevertheless, this matrix is not guaranteed to be positive definite. Then, like [5], we use the approximation proposed by [18], that is $M_t(1 - \bar{\alpha}_t - \bar{\beta}_t)$, where $\bar{\alpha}_t$ and $\bar{\beta}_t$ are two scalars equal to the average value of the elements in the matrix A_t and B_t, respectively, thus ensuring the positive definiteness of the Conditional Covariance matrix.

In empirical studies, only one matrix is parameterized through the Hadamard exponential function: e.g., [5] find that the coefficients of the lagged Realized Covariance matrix are time–varying for the models with a BEKK-type dynamics; for this reason we put $B_t = \beta J_n$.

2.2 Estimation

We estimate the parameters of the proposed models through the Quasi Maximum Likelihood (QML) method in one step. Indeed, we cannot estimate models with a time–varying average level through the covariance targeting procedure,[8] a two-step approach, where at the first step the unconditional covariance matrix is estimated through a consistent estimator based on the sample covariance, that is, $\hat{M} = T^{-1}\sum_{t=1}^{T} C_t$.[9] In the second step, the other parameters are estimated by maximizing the log-likelihood conditional on the estimated unconditional covariance matrix.

Let us assume a Wishart distribution for the Realized Covariance matrix and let $\Phi = \{\phi_i\}$ be the set of unknown parameters; the Quasi log-likelihood (omitting the

[8] See, for example, [6].

[9] Nevertheless, [8] proposed an algorithm that iteratively maximizes the moment-based QML function (Iterative Moment-based Profiling estimator—IMP) thus rendering the estimation feasible for a growing number of assets. However, IMP is less efficient than QML estimator, that maximizes the Quasi log-likelihood in one step.

part that does not depend on Φ) is:

$$LL(\Phi) = \sum_{t=1}^{T} ll_t(\Phi) = -\nu/2 \sum_{t=1}^{T} \left[\ln |S_t| + tr\left(S_t^{-1} C_t\right) \right] \tag{10}$$

Let us consider the score of each parameter ϕ_i:

$$\frac{\partial ll_t(\Phi)}{\partial \phi_i} = -\nu/2 \left[tr\left(S_t^{-1} \frac{\partial S_t}{\partial \phi_i} - C_t S_t^{-1} \frac{\partial S_t}{\partial \phi_i} S_t^{-1} \right) \right] \tag{11}$$

Then, let us compute the conditional expected value of the score:[10]

$$E_{t-1} \frac{\partial ll_t(\Phi)}{\partial \phi_i} = -\nu/2 \left[tr\left(S_t^{-1} \frac{\partial S_t}{\partial \phi_i} - \frac{\partial S_t}{\partial \phi_i} S_t^{-1} \right) \right] = 0 \tag{12}$$

As we can see in Eq. (12), the score is a martingale difference sequence and, by the law of iterated expectations, $E \frac{\partial ll_t(\Phi)}{\partial \phi_i} = 0$. Moreover, the parameter of the Wishart distribution does not influence the other parameter estimates because the first–order conditions are a linear function of ν, then the latter can be set equal to 1 during the estimation process (as in [7]). This is an important result because, even if the assumed distribution is misspecified, the estimator is consistent, given that the conditional expectation of C_t is correctly specified, then its QML interpretation. Consequently, we compute the standard errors through the Sandwich matrix [23].

3 Empirical Analysis

3.1 Dataset

In the empirical analysis, we employ a time series of daily Realized Covariance matrices, from 1 January 2001 to 16 April 2018, of nine assets included in the components of the Dow Jones Industrial Average Index: Apple Inc. (AAPL), Chevron Corporation (CVX), The Walt Disney Company (DIS), The Goldman Sachs Group, Inc. (GS), Home Depot Inc. (HD), International Business Machines Corporation (IBM), Intel Corporation (INTC), 3M Company (MMM), and Exxon Mobil Corporation (XOM). The data we employ is a subset of the dataset used by [5][11] obtained through a grouping algorithm that identifies the assets not having statistically different coefficients: in short words, each group is formed by applying

[10] See [11] for multivariate GARCH models and [6] for CAW models.

[11] Lyudmila Grigoryeva (University of Konstanz) and Juan-Pablo Ortega (University of St.Gallen) provided the dataset for the DJIA companies, while Oleksandra Kukharenko (University of Konstanz) computed the Realized Covariance matrices.

Table 1 Descriptive statistics of realized variances

	Mean	Median	Min.	Max.	St. dev.	Skewness	Kurtosis
AAPL	9.557	4.489	0.203	422.730	16.727	8.045	130.957
CVX	4.197	2.399	0.042	433.818	9.972	22.777	848.737
DIS	5.665	2.595	0.061	240.486	9.851	7.832	120.754
GS	8.040	3.198	0.048	1200.654	30.538	24.060	781.133
HD	5.567	2.700	0.097	346.298	10.353	11.816	298.493
IBM	3.652	1.770	0.023	188.960	7.564	11.038	194.705
INTC	7.389	3.819	0.093	205.325	10.853	5.178	52.038
MMM	3.369	1.796	0.048	405.997	8.549	28.151	1203.749
XOM	4.034	2.139	0.076	444.402	9.946	24.057	937.262

The table reports the Mean, the Median, the Minimum (Min), the Maximum (Max), the Standard Deviation (St. Dev.), the Skewness and the Kurtosis of the Realized Variances. All the variables are expressed in annualized percentage terms. Sample period: 1 January 2001–16 April 2018

the Wald test to the estimated parameters, and then the model is re-estimated by constraining the parameters to be the same within each group.[12]

By looking at the descriptive statistics in Table 1, variances are positively skewed and their empirical distribution has a much higher level of excess of kurtosis than the normal one. Similar comments hold for covariances (for the sake of brevity we do not report the statistics relative to the covariances; they are available on request). In general, variances have a higher average level and show more variability than covariances. Furthermore, from a simple visual inspection (Fig. 1, referred to the AAPL and CVX series), it seems clear that the long–run level of covariances is not constant; in particular it is much higher during periods of market downturns, such as the crisis during the years 2007–2009 or the flash crash on May 6, 2010. Finally, a model with a time–varying average level seems appropriate to offer a better pattern of Realized Covariance series.

3.2 Estimation Results

The estimation period spans 1 January 2001 to 28 March 2017, consisting of 4055 daily observations. Nevertheless, to initialize the MIDAS filter, we need to use the first 264 observations as starting values, so the estimation window reduces to 3791 observations. Indeed, we need 12 lagged values of monthly Realized Covariance matrices, where the latter is the aggregation of 22 daily Realized Covariance matrices.[13] Then, for comparing the estimated models, we remove the

[12] See [5] for an exhaustive description of the grouping algorithm.

[13] As a consequence, the first monthly Realized Covariance is obtained using days from $t = 1$ to $t = 22$, the second one from $t = 23$ to $t = 44$ and so on up to the last period, i.e., from $t = 243$ to $t = 264$.

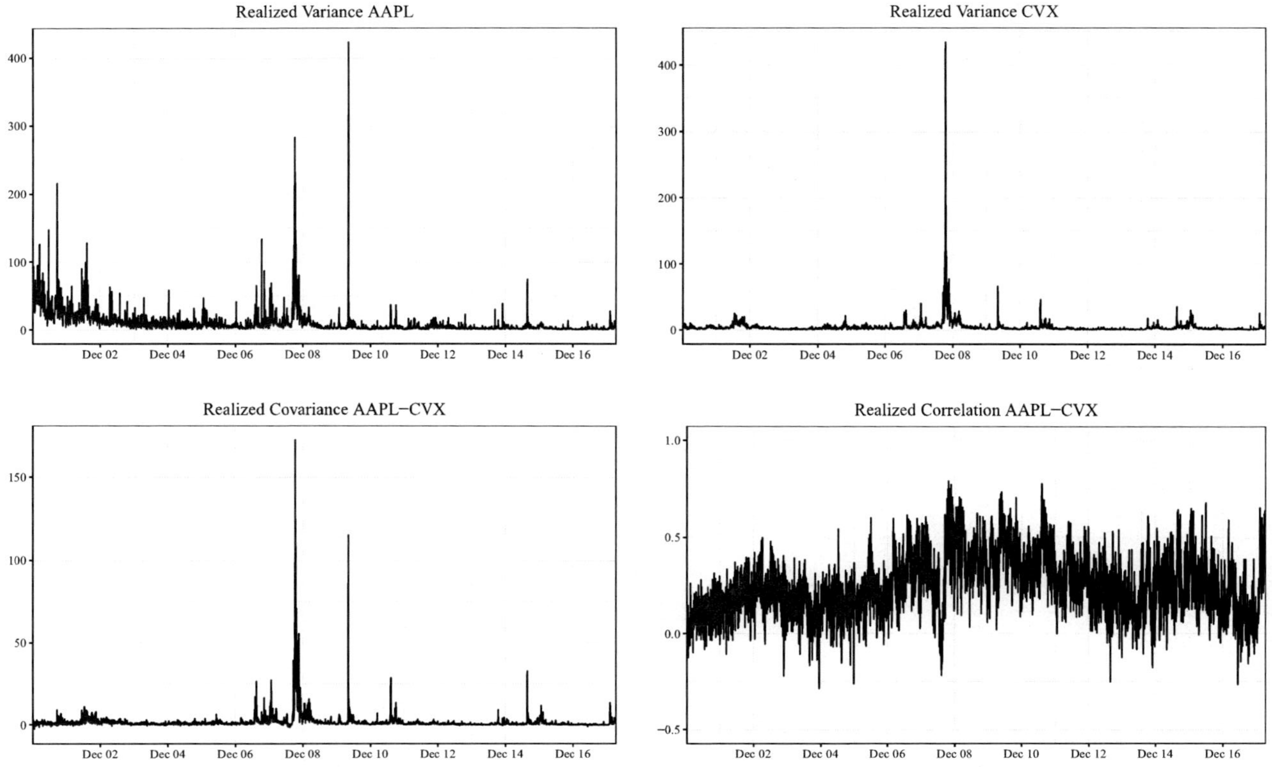

Fig. 1 AAPL and CVX annualized realized variances, covariance and correlation. Sample period: 1 January 2001–16 April 2018

first 264 observations for the scalar ReBEKK, which does not need additional observations. Given the problem of multiple local maxima, the estimation procedure was performed with several starting points, choosing, as final estimates, those with maximum log–likelihood.[14]

The estimates of the coefficients are in line with the previous studies: the estimates of the scalar ReBEKK (see Table 2) imply a very high degree of persistence, that is, the sum of α and β is close to one. Differently, the (co)variance of component models has a persistence lower than the scalar ReBEKK, which is a model assuming a constant long–run component. Relatively to component models, the value of the coefficient θ, representing the impact of the weighted sum of past monthly Realized Covariances on the long–run component, is significant at one percent level, thus supporting the hypothesis of a time–varying average level changing as function of economic conditions. The value of ω_2, the coefficient governing the weighting scheme of past monthly Realized Covariances, is very high and, as a consequence, the monotonically decreasing weights decay very quickly. For what concerns the Hadamard exponential extension of the ReBEKK-MIDAS, the parameter ϕ is significant at 1% in both specifications considering the lagged Realized or the Conditional Correlation, as driving the elements in the matrix A_t. So, the impact coefficient of the lagged Realized Covariance is time–varying and asset–pair specific. In Fig. 2 we show the path of the time-varying coefficient $a_{ij,t}$: it follows the dynamics of the correlation between the two assets, involving a greater weight of the lagged Realized Covariance matrix in high correlation periods. Moreover, when the forcing variable is the Conditional Correlation matrix, rather than the Realized one, the series is smoother, as expected. Finally, adding asset–pair and time–varying coefficients implies a decrease in persistence with respect to alternative models.

3.3 In–Sample Comparison

We evaluate the in–sample performance of the estimated models through the Information Criteria (AIC and BIC), the Quasi Likelihood (QLIKE), a robust statistical loss function in the sense of [22], and the Global Minimum Variance Portfolio (GMVP), an economic loss function [12]. For what concerns the latter, the idea is that a model with superior covariance forecasts should provide a portfolio

[14] More in detail, we used $\alpha = 0.2$ and $\beta = 0.7$, increasing the first (decreasing the second) by 0.02 to get $\alpha + \beta = 0.9$; the sample covariance matrix is the starting value for M in (3) and $\bar{A}$ in (5); $\theta = 0.05$ in increments of 0.01 up to 0.1 and $\omega_2 = 7$ in increments of 1 up to 12; $\phi_A = 0.5$ in increments of 0.05 up to 1.

Table 2 Estimation results of five BEKK–type based models for realized covariance matrices of 9 assets belonging to DIJA

	ReBEKK	MMAReBEKK	ReBEKK-MIDAS	HEReBEKK-MIDAS-P_t	HEReBEKK-MIDAS-R_t
N_p	47	49	49	50	50
α	0.347***	0.351***	0.362***	0.365***	0.393***
	(0.018)	(0.021)	(0.020)	(0.022)	(0.024)
β	0.644***	0.535***	0.545***	0.519***	0.458***
	(0.018)	(0.036)	(0.030)	(0.033)	(0.033)
θ		0.037***	0.040***	0.023***	0.024***
		(0.002)	(0.003)	(0.002)	(0.002)
ω_2		7.102***	9.327***	9.335***	10.809***
		(1.465)	(1.630)	(1.650)	(2.160)
ϕ				0.301***	0.304***
				(0.028)	(0.030)

Dependent variable: Annualized Realized kernel Covariance matrix. Low frequency variable: monthly Realized kernel Covariance. Number of lagged monthly Realized kernel Covariances: k=12. In–sample period: January 28, 2002–March 28, 2017. Number of daily observations: 3791. N_p = Number of estimated parameters. In parentheses, standard errors based on sandwich matrix

The symbol *** indicates that the parameter is significant at a 0.01 level

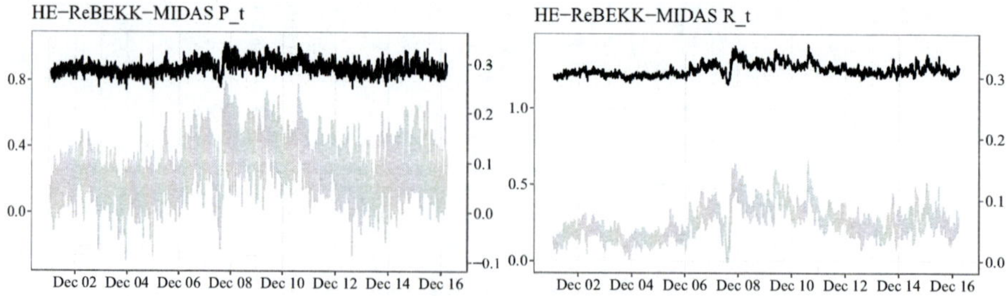

Fig. 2 Estimated $a_{ij,t}$ for the covariance between AAPL and CVX. To the left: $a_{ij,t}$ (black line), realized correlation (gray line). To the right: $a_{ij,t}$ (black line), conditional correlation (gray line). In–sample period: 28 January 2002–28 March 2017

with a lower variance [13]. The loss functions we employ are defined as follows:

$$\begin{aligned} QLIKE &= \sum_{t=1}^{T} tr\left(S_t^{-1} C_t\right) + \ln |C_t| \\ GMVP &= \sum_{t=1}^{T} \hat{w}_t{}' S_t \hat{w}_t \end{aligned} \tag{13}$$

with $\hat{w}_t = S_t^{-1} j \left(j' S_t^{-1} j\right)^{-1}$ and j is a n-dimensional vector of ones.

The Log-Likelihood can be used to compare nested models using the LR test. Among the models considered in Table 3, only ReBEKK–MIDAS is nested in HEReBEKK–MIDAS–P_t and in HEReBEKK–MIDAS–R_t (it can be obtained by the latter by setting $\phi_A = 0$ in Eq. (9)); in these cases the LR test clearly favors the HE specifications because the corresponding statistic is 29.98 and 60.52 respectively (to be compared with the critical value of a chi-squared distribution with 1 degree of freedom). All other models are not nested. In particular, using one of the MIDAS specifications, when $\theta = 0$, the parameter ω_2 is not identified (see the first equation in (5)), so it is a nuisance parameter present only under the alternative hypothesis and non–MIDAS specifications are not nested in it (see [16] and [3]). Moreover, we adopt the exponential weighted moving average (EWMA) of the covariance (also used by [1] among others) and the scalar multivariate Heterogeneous AutoRegressive (vecHAR, [9]) model, estimated via OLS.

Then, we compare the in–sample performance of the estimated models through the Model Confidence Set (MCS) procedure of [20]. To test the null hypothesis of equal predictive capacity we employ the following test statistic:

$$T_{SQ} = \sum_{i \neq j \in \mathcal{M}} \bar{d}_{ij}^2 / \hat{V}ar(\bar{d}_{ij}) \tag{14}$$

where $\bar{d}_{ij}$ is the sample mean difference between the loss function series of models i and j, and $\hat{V}ar(\bar{d}_{ij})$ is the estimated variance of $\bar{d}_{ij}$ through a bootstrap procedure of

Table 3 In–sample performance of the estimated models

	ReBEKK	MMA ReBEKK	ReBEKK MIDAS	HEReBEKK MIDAS-P_t	HEReBEKK MIDAS-R_t	vecHAR	EWMA
Loglik	-34067.35	-34048.49	-34033.76	-34018.77	**−34003.05**		
AIC	17.998	17.989	17.981	17.973	**17.965**		
BIC	18.075	18.069	18.062	18.056	**18.048**		
QLIKE	98.353	98.298	98.254	98.211	**98.167**	98.747	100
Rank	5	4	3	2	1	6	7
p–value	0.000	0.000	0.001	0.012	1.000	0.000	0.000
GMVP	96.313	97.025	95.990	90.192	**89.263**	99.224	100
Rank	3	5	4	2	1	7	6
p-value	0.000	0.000	0.000	0.000	1.000	0.000	0.000

Loglik: maximized value of the log-likelihood. AIC: Akaike Information Criterion. BIC: Bayesian Information Criterion. QLIKE: Quasi-Likelihood function. GMVP: Global Minimum Variance Portfolio. Loglik, AIC and BIC are not reported for EWMA (because it is not estimated) and vecHAR (because it is estimated nonparametrically via OLS). QLIKE and GMVP are expressed as ratios to the benchmark model value (EWMA) multiplied by 100. In bold the best value for each function, for the loss functions the value is with respect to the benchmark. Rank indicates the (inverse) order in which the models are removed in the MCS approach (7 is the first model removed, 1 the best performing model) both for QLIKE and GMVP; p-value is the corresponding p-value of the test statistic. Covariance Proxy: annualized Realized kernel Covariance matrix. In–Sample period: 28 January 2002–28 March 2017
In bold the best value for each function with respect to the benchmark

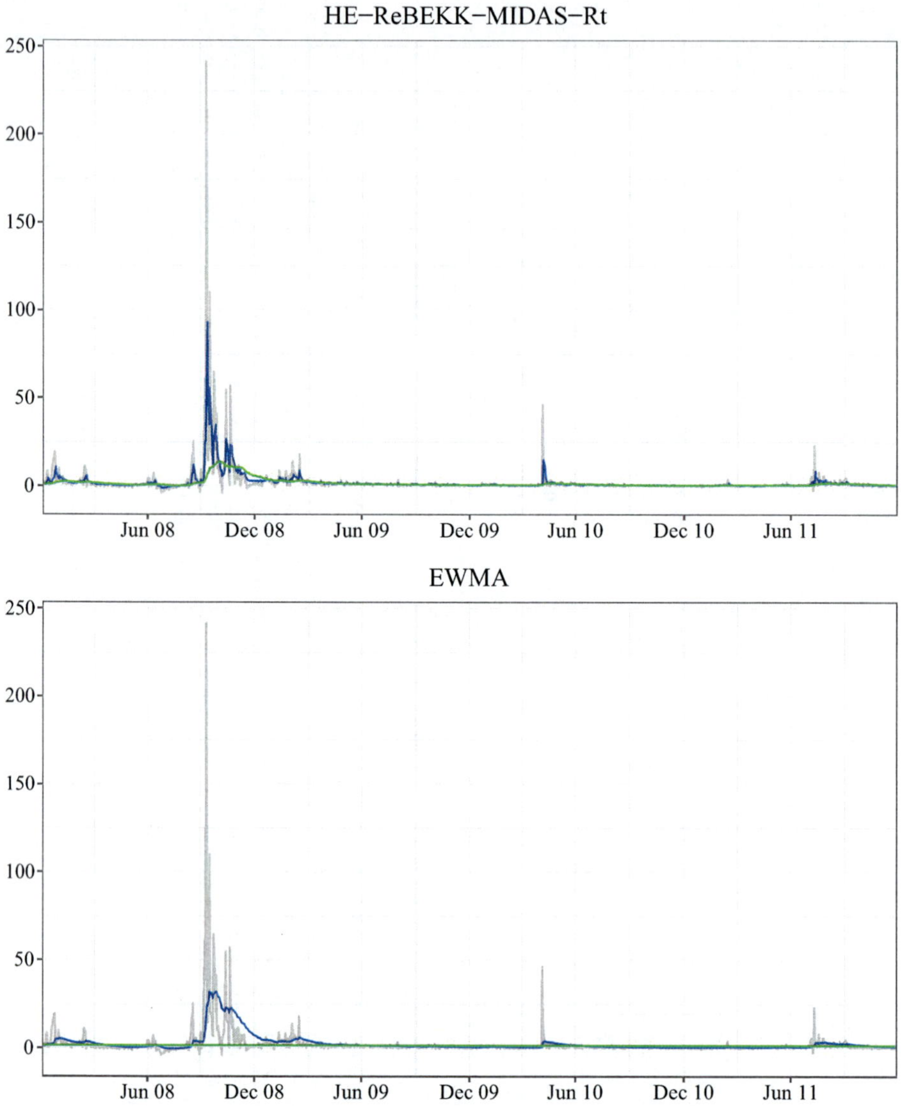

Fig. 3 Results derived from HE–ReBEKK–MIDAS–R_t and EWMA models for the asset pair IBM and XOMM: realized covariance (gray line), conditional covariance (blue line), long-run component (green line). Sample period: January 2, 2008—December 30, 2011

5000 resamples. When the null hypothesis is rejected the worst model is eliminated and the test is repeated for the remaining models until the null hypothesis is not rejected, thus providing the set of models with equivalent predictive capability.

From Table 3 it is clear the superior in–sample forecasting capability of the HEReBEKK–MIDAS–R_t, also supported by the Information Criteria.

In Fig. 3, we provide the dynamics of the Realized Covariance and the estimated Conditional Covariance and long–run component derived from our best model

Table 4 Out–of–sample performance of the estimated models

	ReBEKK	MMA ReBEKK	ReBEKK MIDAS	HEReBEKK MIDAS-P_t	HEReBEKK MIDAS-R_t	vecHAR	EWMA
QLIKE	97.100	96.539	96.507	96.547	**96.387**	98.325	100
Rank	5	3	2	4	1	6	7
p–value	0.000	0.248	0.236	0.233	1.000	0.000	0.000
GMVP	118.640	112.939	109.868	110.088	110.088	140.570	**100**
Rank	6	5	2	3	4	7	1
p–value	0.002	0.517	0.251	0.472	0.720	0.000	1.000

QLIKE: Quasi-Likelihood function. GMVP: Global Minimum Variance Portfolio. QLIKE and GMVP are expressed as ratios to the benchmark model value (EWMA) multiplied by 100. In bold the best value for each function with respect to the benchmark. Rank indicates the (inverse) order in which the models are removed in the MCS approach (7 is the first model removed, 1 the best performing model) both for QLIKE and GMVP; p-value is the corresponding p-value of the test statistic. Covariance Proxy: annualized Realized Covariance. Out–of–Sample period: 29 March 2017–16 April 2018

In bold the best value for each function with respect to the benchmark

(HEReBEKK–MIDAS–R_t) and from the benchmark (EWMA), focusing on the subprime mortgage crisis period (the rest of the series shows a flatter dynamics). An increase in the level of the conditional covariances for both models can be appreciated during the second half of 2008, in correspondence with the subprime mortgage crisis, but the HEReBEKK–MIDAS–R_t follows the turbulent dynamics of this period with several peaks and troughs, while the conditional covariance of EWMA is smooth. Furthermore the long–run component of the former changes according to the abrupt (but not brief) change, while the latter is constant.

3.4 Out–of–Sample Analysis

We conduct also an out–of–sample exercise for the period between 29 March 2017 and 16 April 2018. For this purpose, we generate 264 one-step-ahead forecasts of the covariance matrix, based on the parameters obtained during estimation process. The estimates are updated monthly (with rolling windows of 22 observations), repeating the estimation procedure 12 times.

The out–of–sample performance of the estimated models is evaluated by the QLIKE and GMVP loss functions through the MCS procedure (Table 4). The out-of-sample analysis reveals the best performance of the HEReBEKK-MIDAS-R_t according to the QLIKE. This is a not so obvious result: it is a recurrent feature in empirical applications that the most sophisticated models get a better in–sample fitting, while the simplest models tend to have a better out–of–sample performance [19]. It is also the only model entering the best set in MCS approach if we consider a significance level of the sequential tests of 25%, one of the typical values adopted in such procedure (see, for example, [20] and [7]). Considering lower significance levels, only MIDAS models enter the best set.

For the GMVP, EWMA has the lowest value (consistently with the results of [5]), but its performance is indistinguishable with respect to MIDAS models.

4 Concluding Remarks

In this work, we propose a new class of multivariate component volatility models for Realized Covariance matrices, called ReBEKK-MIDAS, whose estimated coefficients are invariant to the order of the assets and exhibit a computational gain during the estimation process. More in detail, we specify the long–run component as a time–varying intercept, which is a function of the weighted sum (by adopting the MIDAS approach) of the lagged monthly Realized Covariance matrices. Furthermore, we introduce an extension of the basic model, by including the Hadamard exponential function proposed by [5], call it Hadamard Exponential Realized BEKK MIDAS (HEReBEKK-MIDAS). This specification admits asset–pair specific and time–varying impact coefficients, only with one parameter more than the basic model, thus preserving the parsimony of the model.

We have also estimated alternative specifications, not reported in this paper. Like [17] and [7], in this paper we have chosen as long-run variable the monthly Realized Covariance, and a number of lags equal to 12 (one MIDAS year). However, we also estimated the model for two MIDAS years, but we did not have an increase in log-likelihood. We also estimated the model by aggregating the daily Realized Covariances on a quarterly frequency and the estimated results are very similar to our model, but we prefer to rely on the latter in order to preserve the smooth pattern of the long–run component (see [7]). In fact, with a quarterly frequency we have only 4 lags. Finally, we also estimated the multivariate extension of H–MIDAS–CMEM proposed by [21], with very similar results in terms of AIC and BIC compared to our basic model (ReBEKKMIDAS). The results relative to these alternative specifications are available on request.

The proposed model, more specifically that parameterized through the Hadamard exponential function, has the best in–sample performance according to the information criteria, and the MCS approach for the QLIKE and GMVP functions. We evaluate also the out–of–sample forecasting capability of the models through the MCS approach. The forecasting exercise confirms the better performance of the models that allow time–varying and asset–pair specific parameters when we consider the statistical loss function (QLIKE). The out–of–sample performance relative to the economic loss function (GMVP) seems to be satisfactory for the whole class of ReBEKK-MIDAS models.

Future research could consider other drivers of the long–run component, such as macroeconomic variables, to analyze the relationship, in a multivariate framework, between economics and financial volatility. Another issue is related to the number of assets: indeed, for Realized Covariance matrices of large dimensions, we face the *curse of dimensionality problem*. In this work we address this topic to provide the possibility of having different coefficients for each series of covariances in a parsimonious way, by adopting the Hadamard exponential operator, but the matrix of constant parameters of the long–run component remains an increasing function of the number of assets considered. Removing the constant parameter matrix from the QML estimate, for example by adopting the Iterative Moment-based Profiling (IMP) algorithm of [8], the model could be also estimated for high–dimensional Realized Covariance matrices.

Acknowledgments Edoardo Otranto acknowledges financial support from the Italian PRIN 2022 grant "Methodological and computational issues in large-scale time series models for economics and finance" (20223725WE).

References

1. Amendola, A., Braione, M., Candila, V., Storti, G.: A model confidence set approach to the combination of multivariate volatility forecasts. Int. J. Forecast. **36**, 873–891 (2020)
2. Amendola, A., Candila, V., Cipollini, F., Gallo, G.M.: Doubly multiplicative error models with long–and short–run components. Technical Report (2021)
3. Andrews, D.W.K., Cheng, X.: Estimation and inference with weak, semi-strong, and strong identification. Econometrica **80**, 2153–2211 (2012)
4. Bauwens, L., Otranto, E.: Nonlinearities and regimes in conditional correlations with different dynamics. J. Econ. **217**(2), 496–522 (2020)
5. Bauwens, L., Otranto, E.: Modelling realized covariance matrices: a class of Hadamard exponential models. J. Financ. Econ. **21**(4), 1376–1401 (2023)
6. Bauwens, L., Storti, G., Violante, F.: Dynamic conditional correlation models for realized covariance matrices. CORE DP **60** (2012)
7. Bauwens, L., Braione, M., Storti, G.: Forecasting comparison of long term component dynamic models for realized covariance matrices. Ann. Econ. Stat. **123–124**, 103–134 (2016)
8. Bauwens, L., Braione, M., Storti, G.: A dynamic component model for forecasting high-dimensional realized covariance matrices. Econ. Stat. **1**, 40–61 (2017)
9. Chiriac, R., Voev, V.: Modelling and forecasting multivariate realized volatility. J. Appl. Econ. **26**, 922–947 (2011)
10. Colacito, R., Engle, R.F., Ghysels, E.: A component model for dynamic correlations. J. Econ. **164**(1), 45–59 (2011)
11. Comte, F., Lieberman, O.: Asymptotic theory for multivariate GARCH processes. J. Multivariate Anal. **84**(1), 61–84 (2003)
12. Engle, R.F., Colacito, R.: Testing and valuing dynamic correlations for asset allocation. J. Bus. Econ. Stat. **24**(2), 238–253 (2006)
13. Engle, R.F., Kelly, B.: Dynamic equicorrelation. J. Bus. Econ. Stat. **30**(2), 212–228 (2012)
14. Engle, R.F., Kroner, K.F.: Multivariate simultaneous generalized ARCH. Econ. Theory **11**(1), 122–150 (1995)
15. Gallo, G.M., Otranto, E.: Forecasting realized volatility with changing average volatility levels. Int. J. Forecast. **31**, 620–634 (2015)
16. Ghysels, E., Sinko, A., Valkanov, R.: MIDAS regressions: further results and new directions. Econ. Rev. **26**(1), 53–90 (2007)
17. Golosnoy, V., Gribisch, B., Liesenfeld, R.: The conditional autoregressive Wishart model for multivariate stock market volatility. J. Econ. **167**(1), 211–223 (2012)

18. Hafner, C.M., Franses, P.H.: A generalized dynamic conditional correlation model: simulation and application to many assets. Econ. Rev. **28**(6), 612–631 (2009)
19. Hansen, P.R.: A winner's curse for econometric models: on the joint distribution of in-sample fit and out-of-sample fit and its implications for model selection. Technical Report (2010)
20. Hansen, P.R., Lunde, A., Nason, J.M.: The model confidence set. Econometrica **79**(2), 453–497 (2011)
21. Naimoli, A., Storti, G.: Heterogeneous component multiplicative error models for forecasting trading volumes. Int. J. Forecast. **35**, 1332–1355 (2019)
22. Patton, A., Sheppard, K.: Evaluating Volatility and Correlation Forecasts. Handbook of Financial Time Series, pp. 801–838. Springer, Berlin (2009)
23. White, H.: Maximum likelihood estimation of misspecified models. Econometrica **50**(1), 1–25 (1982)

Taxonomy-Based Risk Analysis with a Digital Twin

Giovanni Paolo Sellitto, Tanja Pavleska, Massimiliano Masi, and Helder Aranha

Abstract Risk analysis and risk management are mandatory by law in critical sectors, but also for enterprises, to comply with liability frameworks. However, a gap exists between the establishment of an organizational risk culture and the actual preparedness of organizations. The problem is twofold, as in the ongoing trend towards cyber-physical organizations, it touches both risk awareness and the communication about risks, to mitigate them during operations. The effort to address the risk-culture gaps usually involves a balance of short and long term interventions, to establish a consistent and enterprise-wide risk communication and management strategy. The application of digital tools and enterprise modelling can support the organizations in overcoming this gap.

In this work, we propose framework specifically targeted at reducing the organizational gap between the risk management process and operations, making enterprise risk management usable and effective. This is done by integrating a Digital Twin and visual threat modelling into a goal-based methodology as a means to involve business people in the risk management process, reducing the need for dedicated risk management and security experts.

To illustrate the practical benefits and the viability of this approach, we apply the framework to a simple business case: a Smart Resort operated through a Building Management System.

G. P. Sellitto (✉)
Independent Scholar, Rome, Italy

T. Pavleska
Jozef Stefan Institute, Ljubljana, Slovenia
e-mail: atanja@e5.ijs.si

M. Masi
Independent Scholar, Florence, Italy
e-mail: max@mascanc.net

H. Aranha
Independent Scholar, Lisbon, Portugal

© The Author(s), under exclusive license to Springer Nature Switzerland AG 2023

C. P. Kitsos et al. (eds.), *Statistical Modelling and Risk Analysis*,
Springer Proceedings in Mathematics & Statistics 430,
https://doi.org/10.1007/978-3-031-39864-3_15

Keywords Enterprise risk management · Digital twin · Risk communication

1 Introduction

The International Organization for Standardization (ISO) defines risk [1] as "The effect of uncertainty on objectives". Risk analysis is a framework for decision making under uncertainty. The ISO 31000 "Risk management—Guidelines" provide principles, a framework, and a process for managing risk. This standard helps to identify opportunities and reduce the severity of the consequences of risk on the achievement of organizational objectives. Risk management, as defined by ISO 31000 is an organized process for identifying what may go wrong, quantifying and assessing the associated risks, implementing measures and actions to prevent or treat each identified risk. In the process of enterprise risk management (ERM), it is of paramount importance to plan, identify, evaluate, control and communicate the various aspects related to risk, in order to minimize the negative impact it can have on the organization [2]. Risk analysis typically involves identifying the risks, assessing their probabilities and impacts, ranking them and screening out minor risks and this process requires both business knowledge and experience in dealing with risk. In the same standard, a risk management framework is defined as a "set of components that provide the foundations and organizational arrangements for designing, implementing, monitoring, reviewing and continually improving risk management processes throughout the organization". The importance of communication to build awareness inside organizations exposed to risk requires the adoption of frameworks and instruments that are easy to understand and use by business people, without the necessity for security experts and IT architects. Here the challenge is often represented by the transition from qualitative to quantitative risk management and back and from the (macro) level of business objects to the technical (micro) level of ICT components and services that support enterprise operations.

Usually, business people are able to embrace a qualitative risk management approach, reasoning over business entities and processes. Clearly, this is important for the creation of a "risk culture" in the organization. On the other hand, security experts and risk management standards require a quantitative evaluation of risk and knowledge of the technical components, in order to devise countermeasures and implement them successfully. The difference between the conceptual models used by business people and those used by experts and in the levels of abstraction where the reasoning is rooted often creates a communication mismatch, which in itself can represent a high risk for an organization. As it will become evident from this work, business people and experts reason on different taxonomies[1] to devise

[1] A risk taxonomy is the (typically hierarchical) categorization of risk types. A common approach is to adopt a tree structure, whereby risks higher in the hierarchy are decomposed into their more specific (granular) manifestations.

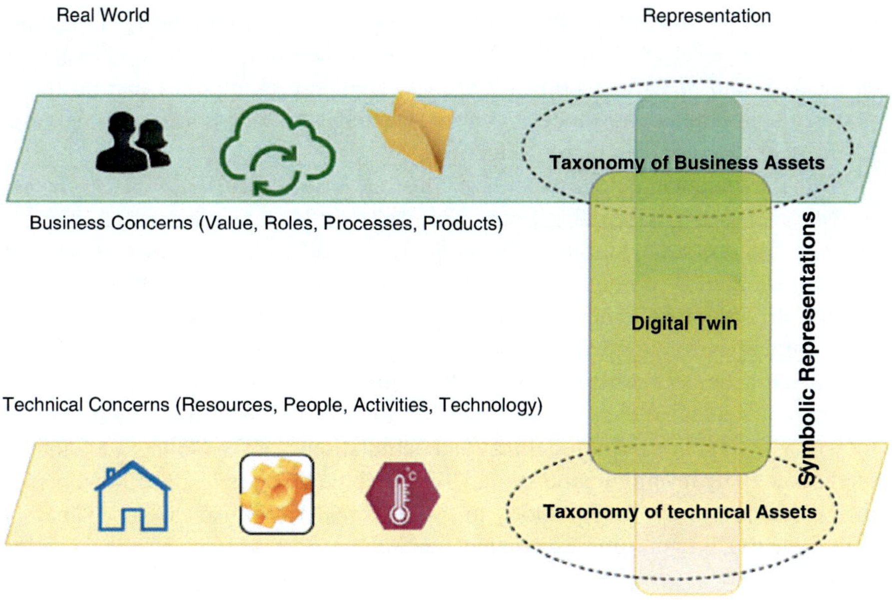

Fig. 1 Bridging business and technical views using a digital twin. This representation is based on a picture by Rémy Fannader, at page 21 of the book *Enterprise architecture fundamentals: using the Pagoda blueprint* [3]. We are thankful to the author for granting us with the right to use a modified version of the original figure

measures and address risk issues (Fig. 1): business people consider Value, Roles, Processes and Products using business-level concepts, while experts and technicians strive to protect Resources, People, Activities and Technology components and solutions, like firewalls, appliances and protocols. In this context, the availability of digital models of the enterprise can help in bridging the two conceptual worlds and to support the communication between business people and experts. In addition, platforms that exploit them for simulations and visual threat modelling create new opportunities both from the point of view of communication and, more in general, for risk management.

Over the last years, Digital Twins have been used to perform simulations on cyber-physical systems to enable security and safety by design. As a results, they have been gaining ground as a useful tool to support the risk management process. There are several definitions for a *Digital Twin*, but in this paper, we rely on the following: a Digital Twin is "a virtual description of a physical product that is accurate to both micro and macro level" [4]. This notion is complemented by the one presented by Eckhart et al. [5], who refers to the Digital Twin as a "virtual replica of the system that accompanies its physical counterpart during its lifecycle, consumes real-time data if required, and has the sufficient fidelity to allow the implementation, testing, and simulation of desired security measures and business continuity plans". In this latter definition, it is clear that the system under consideration can be an organization, as modern organizations are increasingly becoming socio-technical and cyber-physical environments, where usually the human factors pose the most severe risk and weakness points. These types of representations can be used to perform simulations and model-based risk analysis, to assess whether the system is secure or if a recovery plan is effective.

In the remainder of this paper, we illustrate how a Digital Twin can be used in the context of a risk management process to build one or more taxonomies of the assets to protect, which in turn can be used to devise countermeasures through simulations and to mitigate risk. In doing our analysis, we use a Digital Twin to support the risk management process and, at the same time, as a convenient tool for risk communication to non-technical people.

To fulfill the goal outlined above, the paper is organized as follows: In Sect. 2 we present related works; Sect. 3 describes the methodology, illustrating how the tools for Visual Threat Modelling and simulations can be integrated into the methodology by leveraging the Digital Twin and the taxonomies. Section 4 presents the application of the methodology to a simple real world use case and, finally, in Sect. 5 we touch upon future work and conclude.

2 Related Work

Using models for risk analysis and security evaluations is part of a research area named threat modelling. The concepts surrounding the threat modelling process have been defined and conceptualized by the Open Web Application Security Project (OWASP) [6], a framework for identifying and understanding threats. It devises mitigation strategies and best practices on how to protect valuable assets. One of the advantages of using a model-based approach for enterprise risk management is that it allows to quantify its effectiveness, to readjust it as necessary, and to adapt to any organizational change. One example of such application can be found in [7], where authors evaluate the extent to which countermeasures can align with the risk requirements for a company in the Energy sector. Over the past decade, the energy sector has been strongly dedicated to security and risk management, thus many valuable technical works can be found in that research area. The concept of *Digital Twin* and its applications has been the subject of study in many recent works [4, 5]. Digital Twins have been successfully used to perform simulations on cyber-physical systems to enable security and safety by design and, as a result, to support the risk management process [8]. Despite these remarkable technical advances, no potential in the Digital Twin as a tool to support or facilitate inter-stakeholder communication has been pointed out yet, let alone researched at a deeper level. This work is a grass-root contribution in that direction.

In addition to ongoing research, there are also many standardized approaches and legal documents addressing risk management that have been widely used by technical experts and engineers. An exhaustive overview of such work was compiled by the European Union Agency for Cybersecurity—ENISA, in its recent report [9].

For instance, ISO 9001 [10] requires the establishing of a separate risk management practice as part of quality management. NISTIR 8286 [11] establishes a relationship between cybersecurity risk management and ERM, helping cybersecurity risk management practitioners at all levels of the enterprise, in private and public sectors, to better understand and practice cybersecurity risk management in the context of ERM. In other words, it is targeting the technical experts to help them integrate risk management practices in an organization. The approach developed in this work complements the existing frameworks by developing the idea and facilitating the process of inter-stakeholder communication on risk-related issues in the enterprise.

3 Methodology

Enterprise Risk Management aims at reducing risks to acceptable levels and communicating how to address critical issues. In this section, we show how to employ a Digital Twin to perform risk analysis and mitigation until an acceptable security posture is reached, and how to facilitate this process by using visual threat modelling tools. At the core of this process is the progressive refinement of some taxonomies of assets to protect and the use of a Digital Twin to support simulations, risk analysis and risk mitigation. In our case, we used an environment for visual threat modelling to perform quantitative simulations and to devise countermeasures. Once they are implemented, the process is repeated until the risk is reduced to an acceptable level.

3.1 Employing a Digital Twin for Risk Management

The approach is loosely based on the Reference Model for Information Assurance and Security (Fig. 2), RMIAS, which is employed as the backbone of an easy to use framework [12]. In its essence, RMIAS is a goal-based methodology,[2] resembling the Plan-Do-Check-Act (PDCA) cycle as recommended by ISO 27000, tailored for Information Assurance and Security (IAS). Focusing on goals allows security experts to communicate with other stakeholders using concepts that do not require technical knowledge. This generic nature of the model is also what makes it adequate for employment at the core of the ERM process. The objectives and the results of the ERM are meant to be communicated to a variety of stakeholders for whom a strong technical background should not be a prerequisite. Thus, the information must be easy to read for a non-technical person who is not acquainted

[2] Apart from goal-based, there are also threat-based approaches that rely on some detailed knowledge of the threats that can affect business operations or the system lifecycle, derived from previous experiences reported in a threat database.

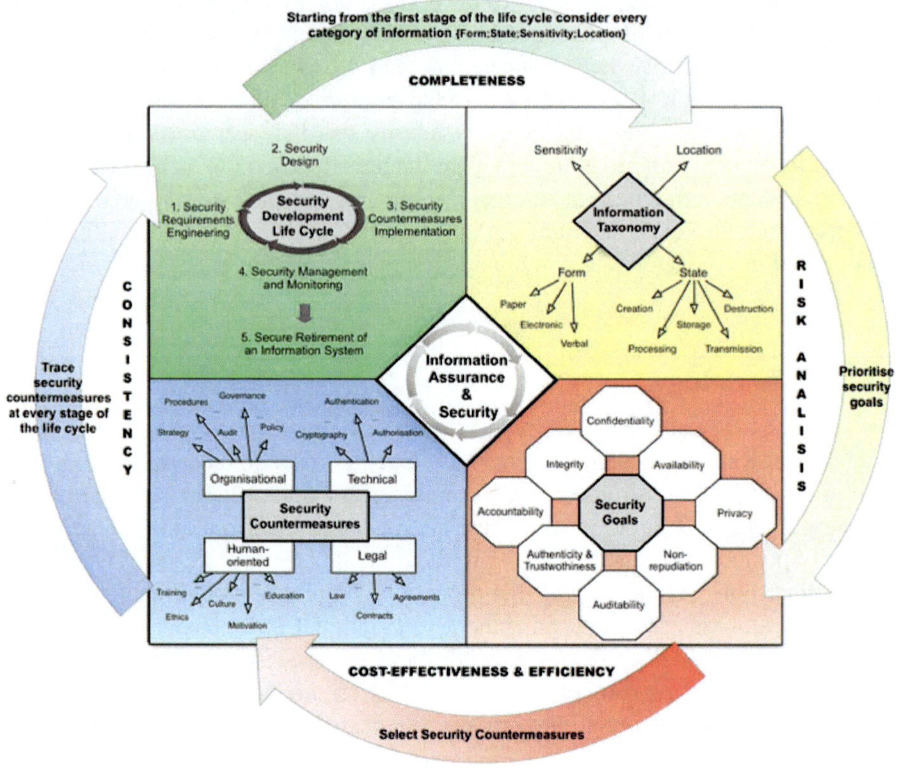

Fig. 2 The reference model for information assurance and security. "A reference model of information assurance and security" (http://RMIAS.cardiff.ac.uk) by Y. Cherdantseva and Hilton is licensed under a creative commons attribution-noncommercial-sharealike 3.0 unported license

with the technical aspects and with the possible solutions to address the potential risk. This is also a major argument for employing a goal-based framework, as it does not require that we provide a detailed threat model straight from the start of our analysis.

Our work extends RMIAS with additional components aimed at facilitating the inter-stakeholder communication. This extended framework is depicted in Fig. 3. At the core of the framework is the digital representation of the enterprise, will all its components (services, processes, people, information, etc.). The ERM process exploits the information from the Digital Twin to devise an adequate risk management strategy. The other parts of the framework are managed through RMIAS to achieve this purpose, coordinating the taxonomies (asset classifications), the security goals and the countermeasures needed for reaching the goals. Finally, the visual threat modelling mediates the process of simulation, testing the relevant scenarios and supporting both the decision-making process and the inter-stakeholder communication. Figure 3 illustrates the components of a methodology compliant with ISO 31000, which is structured through the following steps:

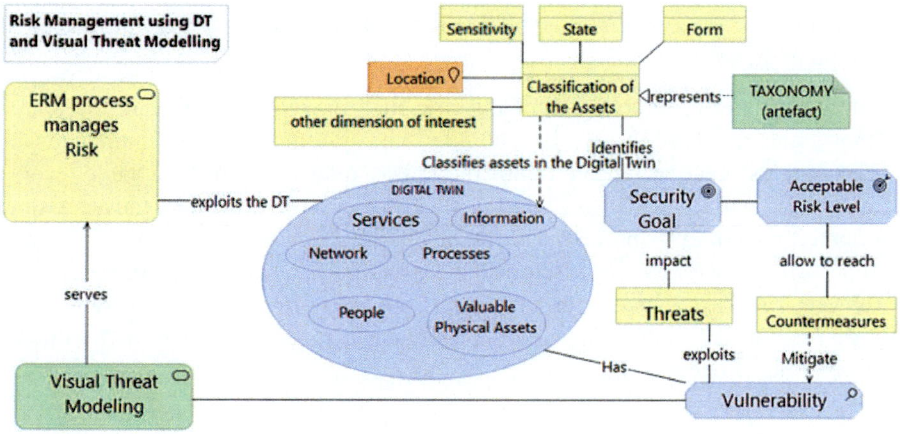

Fig. 3 Risk management using a DT and visual threat modelling

- **Step 1** of the methodology is a stocktaking operation, starting from the business goals, identifying the material and immaterial components (assets) that concur to reach these goals and defining the purpose of each asset from the business point of view. The results of this operation take the form of a taxonomy, where the business objectives are linked to roles, processes, products that concur to the creation of value. In this phase we can also elicit from the stakeholders a first ranking of the elements that may have an impact on business objectives.
- **Step 2** starts from the results of the previous phase and leverages the knowledge provided by the stakeholders and business experts to map the elements that are important business-wise into their physical and technical counterparts, to create a taxonomy of high-value assets. Here the presence of symbolic models of the enterprise can facilitate the proper individuation of the relevant business assets. The ranking of the business assets in this taxonomy, based on their value, also makes evident the level of risk that we can accept for each of them, providing the basis for the subsequent phases.
- In **Step 3**, with the support of the information contained in the Digital Twin, the taxonomy of the assets to be protected is translated by technical people in a taxonomy of the technical components that must be protected, which will be used to perform a quantitative risk analysis. In this step, the use of a Digital Twin as *a virtual description that is accurate to both micro and macro level* facilitates the mapping between the business assets and the corresponding technical components.
- In **Step 4**, the Digital Twin offers a base for quantitative risk analysis and to devise countermeasures. In our case, we used an environment for visual threat

modelling[3] to perform quantitative simulations and to devise the necessary countermeasures.

These four steps perform a complete risk analysis cycle, bringing in the domain knowledge about the threats that can affect the system through its entire lifecycle. The process must be repeated until the risk is reduced to an acceptable level. In this cycle, taxonomies have a central role, as they bridge the qualitative, business oriented risk analysis process and the quantitative and technical risk management approach. Classifying the assets using taxonomies enables us to understand the relevant risks associated with each asset, but also offers a mean to translate the business concerns into a quantitative estimate of risk levels associated with each *item* in the taxonomy.

4 The Smart Resort Business Case

To demonstrate the practical feasibility of our approach, here we apply it to the business case of a resort (Hotel) located in a Smart Building. From the business point of view, the goal of such an enterprise is to provide customers with an enjoyable experience. Whether the enjoyment comes from eating a good meal, relaxing in a luxurious spa, or getting a good night's rest away from home, it is of paramount importance to take care of each individual guest. These objectives are concretely supported by business roles, processes and assets that ensure seamless operations and reduce the risk of events that can have a negative impact on the customers' experience.

From the technical point of view, the operations take place in a location and they are supported by a Building Management System (BMS) that allows to control and operate subsystems like video surveillance, Heat Ventilation and Air Conditioning systems (HVAC), fire detectors, illumination, elevators, water and sewage management. The BMS is usually centrally operated by a Supervisory Control And Data Acquisition (SCADA) system and managed by technical operators in a control room. This is exemplified by the BMS of Fig. 4, operated from a single control room, where a single operator controls HVACs, uninterruptible power supply systems (UPS), the Closed Circuit TV (CCTV) system, and a set of fire alarms.

Historically, these systems were designed with availability in mind, without considering external threats and security countermeasures, since up to the beginning of the twenty-first century they were physically disconnected from the internet. However, with the advent of internet, cloud computing and smart automation, these systems are open to remote connections and exposed to threats that were not considered from the beginning. Examples of these threats are: the possibility to hijack the electrical power control, manipulating HVAC temperature, disrupt sewage control, stealing data from CCTV cameras, activate the fire alarms up to inducing

[3] For this task we used SecuriCAD, which at the time of writing this paper was available for free https://www.foreseeti.com.

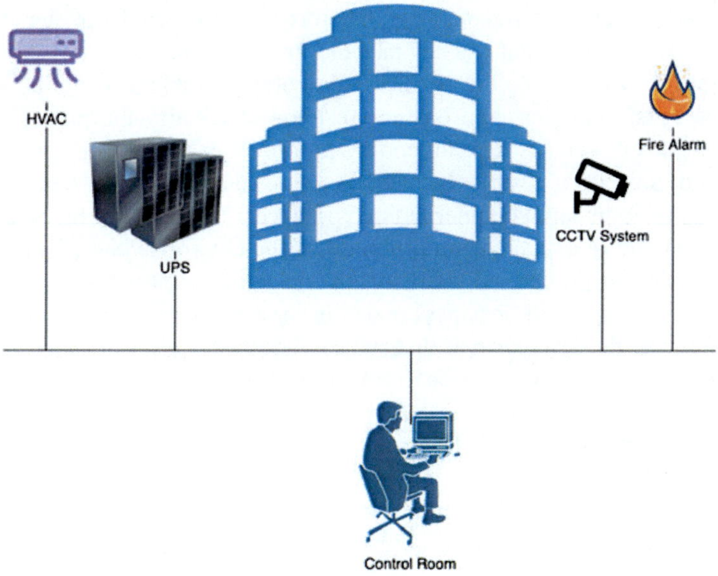

Fig. 4 Schema of a simple BMS for the smart resort

evacuations due to forged fire sensor events. These attacks can be generated either internally (by a disgruntled employee, or by a nefarious actor present in the internal network), or by exploiting remote connections.

To deal with these threats, we must put in place adequate countermeasures. At this point we find the first cognitive mismatch: from the business point of view, a good BMS must be transparent, supporting smooth operations in an adaptive manner, without being visible. This is part of the positive customer's experience. However, from the technical point of view, a BMS is usually composed by devices such as Programmable Logic Controllers and CCTV cameras, which are not easy to reconfigure or patch, and the countermeasures can pose threats to availability (as the system may need a thorough re-certification implying a long downtime). Therefore, the application of cybersecurity countermeasures has to be carefully planned, starting from the evaluation of the effectiveness of each countermeasure, the definition of the budget and planning of the shutdown, to minimize the impact on the building's users. This usually requires a close interaction between technicians, BMS administrators and building managers, to avoid any disruption in the operations. Here we show how the approach presented in the previous section can facilitate this interaction, since the presence of digital models and the possibility to use visual threat modelling can be of help in bridging the conceptual views of business people and technicians and to raise their respective awareness of the value chain and the related risk, agreeing on a common vision. This in turn will help to establish and maintain an enterprise-wide risk culture.

In **Step1**, as described in Sect. 3, we perform the stocktaking operation, starting from the business objectives and linking them to key people, processes, products. This first stocktacking step is performed either by on-site visits or though interviews to the management and stakeholders. The taxonomy produced in step 1 is shared with business management and can take the form of Table 1.

In **Step 2**, we examine the business taxonomy in Table 1 using the knowledge provided by stakeholders and business experts to identify the processes that are important business-wise. The taxonomy produced in this step takes the form of Table 2. In turn, each abstract business process shall be mapped onto its physical, technical and immaterial counterparts, to produce a taxonomy of high-value assets.

The availability of architectural models of the enterprise and of a Digital Twin can facilitate the proper individuation of the relevant business assets. Also, the qualitative ranking of the business assets in the previous taxonomy will provide a base to decide the level of risk that we can accept for each of them. Then, upon elicitation of further information from the stakeholders and through an iterative refinement, we classify the assets according to their impact on the attainment of the overall business objectives.

As a result, we obtain a taxonomy of the high value business assets, which is represented in a simplified form in Table 3. This table represents only an excerpt of the complete taxonomy, since we considered only some high value components.

Table 1 First taxonomy of business goals (Step 1)

Objective	Key components	Impact	Score
provide customers with an enjoyable experience	personnel's kindness	HIGH	95
	proper temperature in the building	HIGH	80
	proper humidity in the building	HIGH	80
	comfortable illumination in the building	HIGH	70
	avoid incidents caused by fire	HIGH	75
	avoid physical intrusion of thieves	HIGH	90
	avoid disruption of the elevators	MEDIUM	50
	visible corporate identity and brand	MEDIUM	40
	broadcast relaxing music	LOW	30
	healthy, tasty and abundant food	LOW	25
	take care of customer's feedback	LOW	20
	clean rooms and shared spaces	HIGH	75

Table 2 Mapping between business goals and processes (Step 2.1)

Key components	Impact	Score	Business process
personnel's kindness	HIGH	95	Customer Care
proper temperature in the building	HIGH	80	Climate Management
proper humidity in the building	HIGH	80	Climate Management
comfortable illumination	HIGH	70	Lighting Maintenance
avoid incidents caused by fire	HIGH	75	Fire Safety Management
avoid physical intrusion of thieves	HIGH	90	Building Surveillance
avoid disruption of the elevators	MEDIUM	50	Lift Maintenance
visible corporate identity and brand	MEDIUM	40	Branding Management
broadcast relaxing music	LOW	30	Customer Care
healthy, tasty and abundant food	LOW	25	Restaurant
take care of customer's feedback	LOW	20	Customer Care
clean rooms and shared spaces	HIGH	75	Customer Care

Table 3 Taxonomy of high value business assets (excerpt) (Step 2.2)

Key components	Impact	Score	Business process
personnel's kindness	HIGH	95	Customer Care
avoid physical intrusion of thieves	HIGH	90	Building Surveillance
proper temperature in the building	HIGH	80	Climate Management
proper humidity in the building	HIGH	80	Climate Management
avoid incidents caused by fire	HIGH	75	Fire Safety Management
clean rooms and shared spaces	HIGH	75	Customer Care

in the form of business processes, while we should also consider people, products and other material and immaterial assets, like intellectual property, buildings, environmental resources that have an impact on the value chain. This approach, in which a business-driven risk analysis is performed in the first step and then only the high-impact components go under further evaluation, is suggested by ANSI/ISA [13].

In **Step 3**, with the support of the simplified architectural model represented in Fig. 4, the taxonomy of the high value assets is translated by technical people in a taxonomy of the (technical) components that must be protected, represented in Table 4.

For the sake of simplicity, here we consider only those components that are part of the Building Management System, namely Building Surveillance, Climate Management and Fire Safety Management. For these processes, the inclusiveness of the proposed approach and the insights gained through the availability of digital tools in the context of a well structured risk management framework are valuable in view of the establishment of an enterprise-wide risk culture. This is especially true in comparison with the classic solutions, where the risk evaluation and treatment activities are performed by technical experts, often external to the organization. Based on Table 4 and using the Digital Twin, we can produce a more detailed taxonomy, comprising the physical components and devices that form the assets, but also messages, data and functionalities that must be protected. In this taxonomy, represented in Table 5, some new processes are added, aggregating common functionalities like those pertaining to Remote Control.

Finally, in **Step 4**, based on the taxonomy represented in Table 5, we obtain a representation of the assets in a Meta Attack Language (MAL), which is the

Table 4 Taxonomy of technical assets, limited to BMS (Step 3)

Impact	Score	Business process	Key assets
HIGH	90	Building Surveillance	CCTV System
HIGH	90	Building Surveillance	Control Room
HIGH	80	Climate Management	HVAC System
HIGH	80	Climate Management	Control Room
HIGH	80	Fire Safety Management	Control Room
HIGH	75	Fire Safety Management	Fire Alarms

Table 5 Taxonomy of physical assets, limited to BMS (Step 3)

Score	Business process	Key assets	Technical Component
90	Building Surveillance	CCTV System	Data
90	Building Surveillance	CCTV System	Cameras
90	Building Surveillance	CCTV System	UPS
80	Climate Management	HVAC	Actuators
80	Climate Management	HVAC	Temperature Controller
80	Climate Management	HVAC	Sensors Devices
80	Climate Management	HVAC	Control Room
75	Fire Safety Management	Fire Alarms	Sensor Data
75	Fire Safety Management	Fire Alarms	Modem/TCP
75	Remote Control	Control Room	SCADA Server
75	Remote Control	Control Room	Console

representation used in SecuriCAD for visual threat modelling. To achieve the required level of detail, an enriched version of the previous table is produced, in the form of Table 6, containing some additional information about (a) the location of the items in a reference conceptual space (the Reference Architectural Model for Industrie 4.0, RAMI 4.0 [14]) and (b) the relevant controls that should be performed on those items, derived from the guidelines published by the National Institute of Standards and Technology (NIST).[4] These pieces of information are part of the Digital Twin. In order to perform a business-oriented security reasoning, we carry out first some attack scenarios which could have higher business impact, if successful. In case of a new system, with no history of previous threats, we can rely on some databases of canonical attacks.

For an existing system, we can perform risk analysis and borrow from the attack strategies for Business Management Systems listed in knowledge bases, like [16]. Additional sources can be used to assess cybersecurity risks and specific attack scenarios for a particular context. Once the attack scenarios are identified, the attack simulations are run with SecuriCAD. The outcome of the simulation is the likelihood of attack success for each scenario of interest, as represented by the third column in Table 7. A business decision must then be made on whether the modelled security posture is cost-effective and acceptable for the given scenario. If not, we shall apply some countermeasures and re-run the simulations on the new configuration, until an acceptable solution is found for all the attack scenarios, e.g.reaching the situation illustrated the last column in Table 7. The devised countermeasures are then transferred to the real system.

In order to provide a practical example of some typical attack scenarios, Fig. 5 reports an attack directed to hijack the electrical power control and one aimed at manipulating the temperature in the resort. In both, the attacker leverages some not previously known vulnerabilities of the system (the so called "zero day vulnerabilities").

---

[4] Using the NIST framework in conjunction with RAMI 4.0 is detailed in [15].

Table 6 Excerpt of the taxonomy formed in Step 3

Name	RAMI Coordinate	NIST	MAL
HVAC	Asset / FieldDevice / Inst Prod	ID.AM-1	Actuator (Component)
Fire Sensor	Asset / FieldDevice / Inst Prod	ID.AM-1	Sensor (Component)
SCADA Server	Asset / Station / Inst Prod	ID.AM-1	ControlServer (Host)
UPS	Asset / FieldDevice / InstProd	ID.AM-1	Sensor
Integration Layer (empty)			
Modbus/TCP	Comm / ControlDevice / Inst Prod	ID.AM-2	ConnectionRule Level2 (DataFlow)
Control Messages	Information / FieldDevice / Inst Prod	ID.AM-2	ICSControlData (Data)
Data Read From Sensors	Information / FieldDevice / Inst Prod	ID.AM-2	ICSControlData (Data)
Regulate the HVAC	Functional / ControlDevice / Inst Prod	ID.AM-3	Important assets
Change temperature	Business / ControlDevice / Inst Prod	ID.AM-3	Important assets

Table 7 Two attack scenarios in a building management system

Attack ID	Description	Risk level before mitigation	Mitigation	Risk Level after mitigation
BMS1	The attacker is impersonating the SCADA user: without any protection, the attacker can have full access to the PLC that manages the UPS.	100%	Add a Multi Factor Authentication to reduce the risk of an attacker impersonating users.	39%
BMS2	The attacker compromises the SCADA and then the PLC of the HVACs: without any protection, the attacker can exploit the full network access of the workstation, scan the network, access the PLC and exploit the vulnerabilities.	100%	Add a logical segregation and an Intrusion Detection System, to limit and discover the movements of an attacker.	29%

In a first scenario (BMS1), the attacker compromises the credentials of the SCADA operator (*SCADA User*), while in the second (BMS2) he gains access to the *Network*, compromising the SCADA. Performing a first round of simulations with SecuriCAD, the likelihood of success for this attack is 100% (third column of Table 7). This is due to the fact that the system was not designed for cybersecurity: any attack would result in reaching the target. However, when we add some countermeasures required by technical guidelines, such as a Multi Factor Authentication for the SCADA operator or a logical separation in the network using Virtual Local Area Networks (VLANs) and an Intrusion Detection System, the attack becomes more difficult. Performing a second round of SecuriCAD simulation, the likelihood of success decreases respectively to 39 and 29%, which were considered acceptable from the business point of view. These values, obtained from simulations after the introduction of countermeasures, are reported in the last column of Table 7.

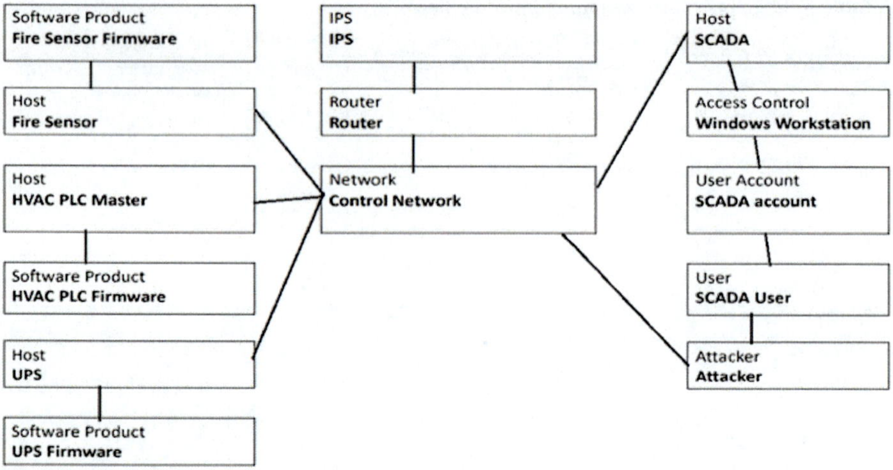

Fig. 5 Visual threat modelling (Step 4)

5 Conclusion and Future Work

In this paper, we illustrated how taxonomies, digital enterprise models and visual modelling tools can be used to perform risk analysis and risk management. Moreover, we leveraged these tools to bridge the communication gap between technical and non-technical people in order to bring all relevant stakeholders into the decision-making process related to enterprise risk management. The presence of a Digital Twin enables the modelling of threats taking into account assets, data flows and messages, as well as functionalities and business objectives. Moreover, the use of visual threat modelling tools and formal languages enables simulation of the potential threats, attacks, and vulnerabilities, further facilitating the decision-making processes related to risk management.

As a future work, we aim at formalizing the approach to allow use of the taxonomies for automatic detection of the countermeasures, making the risk management more independent by the presence of technical experts and further facilitating the inter-stakeholder communication.

References

1. ISO 31000:2018 Risk management — Guidelines. ISO - International Organization for Standardization, Geneva (2018). https://www.iso.org/iso-31000-risk-management.html
2. ISO 31000:2018 Risk management — a practical guide. ISO - International Organization for Standardization, Geneva (2021). https://www.iso.org/publication/PUB100464.html
3. Fannader, R.: Enterprise Architecture Fundamentals: Using the Pagoda Blueprint. Izzard Ink Publishing, Salt Lake City (2021)

4. Grieves, M.: Digital Twin: Manufacturing Excellence Through Virtual Factory Replication, LLC Dassault Systemes, DELMIA Resource Center, Velizy-Villacoublay (2015)
5. Eckhart, M., Ekelhart, A.: In: Biffl, S., Eckhart, M., Luder, A., Weippl, E. (eds.) Digital Twins for Cyber-Physical Systems Security: State of the Art and Outlook, pp. 383–412. Springer, Cham (2019). https://doi.org/10.1007/978-3-030-25312-7 14
6. OWASP security knowledge framework (2021). https://owasp.org/www-project-security-knowledge-framework
7. Korman, M., Valja, M., Bjorkman, G., Ekstedt, M., Vernotte, A., Lagerstrom, R.: Analyzing the effectiveness of attack countermeasures in a SCADA system. In: Proceedings of the 2nd Workshop on Cyber-Physical Security and Resilience in Smart Grids, SPSR-SG@CPSWeek 2017, pp. 73–78. ACM, Pittsburg (2017). https://doi.org/10.1145/3055386.3055393
8. Jones, D., Snider, C., Nassehi, A., Yon, J., Hicks, B.: Characterising the digital twin: a systematic literature review. In: CIRP Journal of Manufacturing Science and Technology, vol. 29, part A, pp. 36–52 (2020). https://doi.org/10.1016/j.cirpj.2020.02.002
9. ENISA: risk management standards - analysis of standardisation requirements in support of cybersecurity policy. The European Union Agency for Cybersecurity, Attiki (2022). https://doi.org/10.2824/001991
10. ISO 9001:2015 quality management systems — requirements. Standard, ISO- International Organization for Standardization, Geneva (2015). https://www.iso.org/standard/62085.html
11. Stine, K., Quinn, S., Witte, G., Gardner, R.: NISTIR 8286 integrating cybersecurity and enterprise risk management (ERM), Gaithersburg, (2020). https://doi.org/10.6028/NIST.IR.8286
12. Cherdantseva, Y., Hilton, J.: A reference model of information assurance and security. In: 2013 International Conference on Availability, Reliability and Security, pp. 546–555. IEEE, Piscataway (2013). https://doi.org/10.1109/ARES.2013.72
13. ANSI/ISA-62443-3-2-2020 Security for industrial automation and control systems, Part 3-2: Security risk assessment for system design. Standard, International Society for Automation, 3252 S. Miami Blvd. 102 Durham, 27703 (2020)
14. Schweichhart, K.: Reference Architectural Model Industrie 4.0 (RAMI 4.0) (2015)
15. Pavleska, T., Aranha, H., Masi, M., Sellitto, G.P.: Drafting a cybersecurity framework profile for smart grids in EU: a goal-based methodology. In: Bernardi, S., Vittorini, V., Flammini, F., Nardone, R., Marrone, S., Adler, R., Schneider, D., Schleiß, P., Nostro, N., Olsen, R.L., Salle, A.D., Masci, P. (eds.) Dependable Computing - EDCC 2020 Workshops - AI4RAILS, DREAMS, DSOGRI, SERENE 2020, Munich, Germany, September 7, 2020, Proceedings. Communications in Computer and Information Science, vol. 1279, pp. 143–155. Springer, Cham (2020). https://doi.org/10.1007/978-3-030-58462-7 12
16. The MITRE corporation: MITRE ATT&CK Matrix for Enterprise (2022). https://attack.mitre.org/

Advanced Lattice Rules for Multidimensional Sensitivity Analysis in Air Pollution Modelling

Venelin Todorov and Ivan Dimov

Abstract Sensitivity analysis is an advanced and efficient technique for verification and improvement of mathematical models. Advanced stochastic approaches based on lattice rules with optimal generating vectors will be applied for sensitivity analysis of large-scale air pollution model. The obtained lattice rule with our optimal generating vector has an optimal rate of convergence for the corresponding class of functions which define the sensitivity indices in the multidimensional air pollution model. The constructed lattice rule improves the results by the other lattice rules and the modified Sobol sequence and gives sufficient accuracy for most of the sensitivity indices.

Keywords Monte Carlo methods · Sensitivity analysis · Multidimensional integrals · Air pollution modelling

1 Introduction

By definition Sensitivity Analysis (SA) is the study of how much the uncertainty in the input data of a model (due to any reason: inaccurate measurements or calculation, approximation, data compression, etc.) is reflected in the accuracy of the output results [6, 11]. SA is very important in today's world when modelling often is the main tool to investigate a complex phenomenon [7, 9, 14, 16, 18]. The main

V. Todorov (✉)
Institute of Mathematics and Informatics, Bulgarian Academy of Sciences, Sofia, Bulgaria

Institute of Information and Communication Technologies, Bulgarian Academy of Sciences, Sofia, Bulgaria
e-mail: vtodorov@math.bas.bg; venelin@parallel.bas.bg

I. Dimov
Institute of Information and Communication Technologies, Bulgarian Academy of Sciences, Sofia, Bulgaria
e-mail: ivdimov@bas.bg

© The Author(s), under exclusive license to Springer Nature Switzerland AG 2023
C. P. Kitsos et al. (eds.), *Statistical Modelling and Risk Analysis*, Springer Proceedings in Mathematics & Statistics 430,
https://doi.org/10.1007/978-3-031-39864-3_16

problem in SA is the evaluation of total sensitivity indices (SIs). The mathematical formulation of estimating SIs is represented by a set of multidimensional integrals (MIs) [2, 5]. Monte Carlo (MC) methods are the best tool to solve MIs [1, 2, 12].

It is assumed that the mathematical model can be presented by a model function

$$u = f(x), \quad \text{where} \quad x = (x_1, x_2, \ldots, x_s) \in U^s \equiv [0; 1]^s \tag{1}$$

is a vector of input parameters with a joint **p**robability **d**ensity **f**unction (p.d.f.) $p(x) = p(x_1, \ldots, x_s)$.

The concept of Sobol approach is based on a decomposition of an integrable model function f into terms of increasing dimensionality [13]:

$$f(x) = f_0 + \sum_{\nu=1}^{s} \sum_{l_1<\ldots<l_\nu} f_{l_1\ldots l_\nu}(x_{l_1}, x_{l_2}, \ldots, x_{l_\nu}), \tag{2}$$

where f_0 is a constant. The representation (2) is referred to as the ANOVA-representation of the model function $f(x)$ if each term is chosen to satisfy the following condition [13]:

$$\int_0^1 f_{l_1\ldots l_\nu}(x_{l_1}, x_{l_2}, \ldots, x_{l_\nu})dx_{l_k} = 0, \quad 1 \le k \le \nu, \quad \nu = 1, \ldots, s.$$

It guarantees that the functions in the right-hand side of (2) are defined in a unique way, where $f_0 = \int_{U^s} f(x)dx$. The quantities

$$\mathbf{D} = \int_{U^s} f^2(x)dx - f_0^2, \quad \mathbf{D}_{l_1\ \ldots\ l_\nu} = \int f_{l_1\ \ldots\ l_\nu}^2 dx_{l_1} \ldots dx_{l_\nu} \tag{3}$$

are the so-called total and partial variances, respectively. A similar decomposition holds for the total variance that is represented by the corresponding partial variances: $\mathbf{D} = \sum_{\nu=1}^{s} \sum_{l_1<\ldots<l_\nu} \mathbf{D}_{l_1\ldots l_\nu}$. The main sensitivity measures following the Sobol approach are the so-called Sobol global sensitivity indices [11, 13] defined by

$$S_{l_1\ \ldots\ l_\nu} = \frac{\mathbf{D}_{l_1\ \ldots\ l_\nu}}{\mathbf{D}}, \quad \nu \in \{1, \ldots, s\}. \tag{4}$$

and the **t**otal sensitivity **i**ndex (TSI) of an input parameter x_i, $i \in \{1, \ldots, s\}$ defined by [11, 13]:

$$S_i^{tot} = S_i + \sum_{l_1 \neq i} S_{il_1} + \sum_{l_1, l_2 \neq i, l_1<l_2} S_{il_1 l_2} + \ldots + S_{il_1\ldots l_{s-1}}, \tag{5}$$

where S_i is called *the main effect (first-order sensitivity index)* of x_i and $S_{il_1 \ldots l_{j-1}}$ is the $j^{\text{-th}}$ order sensitivity index. The higher-order terms describe the interaction effects between the unknown input parameters $x_{i_1}, \ldots, x_{i_\nu}, \nu \in \{2, \ldots, s\}$ on the output variance. It means that the mathematical treatment of the problem of providing global SA consists in estimating total sensitivity indices (5) of corresponding order that, based on the formulas (3) and (4), transforms to computing MIs.

2 The Stochastic Approaches

The first algorithm that we are going to use is the modified Sobol sequence based on procedure of shaking [5].

We will use a method that looks like the stratified symmetrised MC [12]. For our MC algorithm based on modified Sobol sequence (MCA-MSS-2-S) [4], the original domain of integration is divided into m^s disjoint subdomains with equal volumes $\mathbf{K}_i^s, i = 1, \ldots, m^s$, where m is the number of subintervals used for the partition in each dimension. Here two pseudorandom points are generated. The first ξ_i is generated uniformly distributed inside the subdomain $\mathbf{K}_i^s$ and second $\xi_i^{'}$ is computed to be symmetric to ξ_i according to the central point s_i in $\mathbf{K}_i^s$. The concept in two-dimensional case is illustrated on Fig. 1.

The value of the integral can be approximated [3]:

$$I(f) \approx \frac{1}{2m^s} \sum_{i=1}^{m^s} \left[f(\xi_i) + f(\xi_i^{'}) \right].$$

It is proved that the algorithm MCA-MSS-2-S has an optimal rate of convergence $(n^{-\frac{1}{2}-\frac{2}{s}})$ for the class of continuous functions with continuous first derivatives and bounded second derivatives [4]. The reason to choose this method between the other

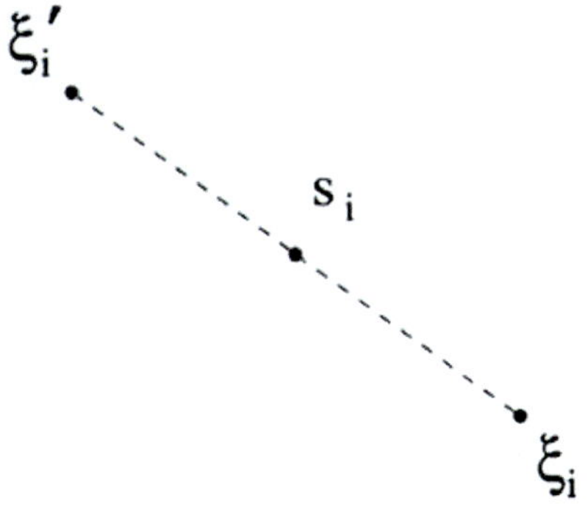

Fig. 1 Generation of a pseudorandom point $\xi_i (\xi_i^{'}) \in \mathbf{E}_i^2$

modified Sobol algorithms is the lowest computational complexity [3] and faster computational time comparable with the other methods in our study.

Now we will use the so called lattice sequences (LS). To introduce rank-1 LS we will use the following formula [15]:

$$\mathbf{x}_k = \left\{ \frac{k}{N} \mathbf{z} \right\}, \; k = 1, \ldots, N, \tag{6}$$

where N is an integer, $N \geq 2$, $\mathbf{z} = (z_1, z_2, \ldots z_s)$ is the generating vector and $\{z\}$ denotes the fractional part of z. For the definition of the $E_s^{\alpha}(c)$ and $P_{\alpha}(z, N)$ see [15].

In 1959 Bahvalov proved that there exists an optimal choice of the generating vector $\mathbf{z}$:

$$\left| \frac{1}{N} \sum_{k=1}^{N} f\left(\left\{\frac{k}{N}\mathbf{z}\right\}\right) - \int\limits_{[0,1)^s} f(u)du \right| \leq cu(s, \alpha) \frac{(\log N)^{\beta(s,\alpha)}}{N^{\alpha}}, \tag{7}$$

for the function $f \in E_s^{\alpha}(c)$, $\alpha > 1$ and $u(s, \alpha)$, $\beta(s, \alpha)$ do not depend on N.

The generating vector $\mathbf{z}$ which satisfies (7), is an optimal generating vector [15] and the main difficulty lies in the construction of the optimal vectors for very high dimensions.

The first generating vector in construction of our LS is the generalized Fibonacci numbers of the corresponding dimension and the method will be called L-FIB [15]. L-FIB will use the following generating vector [15]:

$$\mathbf{z} = (1, F_n^{(s)}(2), \ldots, F_n^{(s)}(s)), \tag{8}$$

where we use that $F_n^{(s)}(j) := F_{n+j-1}^{(s)} - \sum_{i=0}^{j-2} F_{n+i}^{(s)}$ and $F_{n+l}^{(s)}$ $(l = 0, \ldots, j-1, j$ is an integer, $2 \leq j \leq s)$ is the term of the s-dimensional Fibonacci sequence [15].

The second LS is a standard bijectional lattice sequence (L-BIJ) which applies the polynomial transformation function

$$\varphi(t) = 3t^2 - 2t^3$$

to a nonperiodic integrand to make it suitable for applying a lattice rule [8]. The transformation must satisfy the following conditions

$$\varphi(0) = 0, \; \varphi(1) = 1, \; \varphi'(t) > 0.$$

Thus φ is a continuous bijection from [0, 1] to [0, 1].

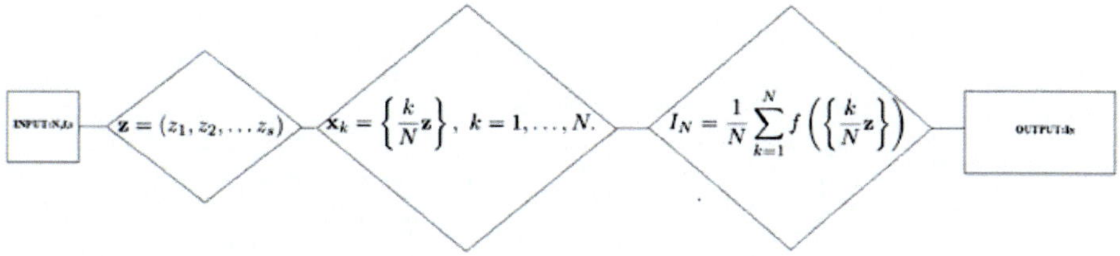

Fig. 2 The flowchart of the algorithm

Now we will consider a special LS based on the component by component construction method (CBCCM) [1, 10]. Applying CBCCM we obtain an optimal generating vector based on the rank-1 lattice rules with prime number of points and with product weights. For this method we will use the notation L-PROD. At the first step of L-PROD the s dimensional optimal generating vector $\mathbf{z} = (z_1, z_2, \ldots z_s)$ is generated by the CBCCM. The second step of the algorithm include generating the points of lattice rule by formula $\mathbf{x}_k = \left\{\frac{k}{N}\mathbf{z}\right\}, \; k = 1, \ldots, N$. And at the third and last step of L-PROD an approximate value I_N of the MI is evaluated by the formula:

$$I_N = \frac{1}{N}\sum_{k=1}^{N} f\left(\left\{\frac{k}{N}\mathbf{z}\right\}\right).$$

The algorithm has an optimal rate of convergence [10, 15]:

$$D_N^* = \mathcal{O}\left(\frac{log^s N}{N}\right).$$

The steps of working for the method are given on the flowchart on Fig. 2.

Since this is a lattice rule, the computational complexity is linear and this directly leads from the flowchart of the algorithm presented on the above figure. Indeed, the advantage of the lattice method is the linear computational complexity and reduced time for calculating the multidimensional integrals. The number of calculation required to obtain the generating vector is asymptotically less than $O(N)$. The generation of a new point requires constant number of operations thus to obtain a lattice set of the described kind consisting N points, $O(N)$ number of operations are necessary.

3 Case Study: UNI-DEM Model

The input data for SA has been obtained during runs of a large-scale mathematical model for remote transport of air pollutants—**Unified Danish Eulerian Model** (UNI-DEM). UNI-DEM is described mathematically [17, 19, 20]) by the following system of partial differential equations:

$$\frac{\partial c_s}{\partial t} = -\frac{\partial(uc_s)}{\partial x} - \frac{\partial(vc_s)}{\partial y} - \frac{\partial(wc_s)}{\partial z} +$$
$$+\frac{\partial}{\partial x}\left(K_x\frac{\partial c_s}{\partial x}\right) + \frac{\partial}{\partial y}\left(K_y\frac{\partial c_s}{\partial y}\right) + \frac{\partial}{\partial z}\left(K_z\frac{\partial c_s}{\partial z}\right) +$$
$$+E_s + Q_s(c_1, c_2, \ldots, c_q) - (k_{1s} + k_{2s})c_s, \quad s = 1, 2, \ldots, q.$$

The number of q equations in this system is equal to the number of pollutants studied by the model. The other dimensions included in the model are described below:

c_s—pollutant concentrations,
u, v, w—wind components along the coordinate axes,
K_x, K_y, K_z—diffusion coefficients,
E_s—space emissions,
k_{1s}, k_{2s}—dry and wet deposit coefficients, respectively ($s = 1, \ldots, q$),
$Q_s(c_1, c_2, \ldots, c_q)$—nonlinear functions describing chemical reactions between pollutants.

Short description of the input parameters and pollutants setup from the perspective of Bulgaria and Europe can be found in the [21]. The computational experiments are conducted using the MATLAB environment. The tests described in the paper are performed on a user laptop with 6-core processor and 16 GB RAM.

3.1 *Sensitivity Studies with Respect to Emission Levels*

Firstly we will study the sensitivity of the model output (in terms of mean monthly concentrations of several important pollutants) with respect to variation of input emissions of the anthropogenic pollutants. The anthropogenic emissions input consist of 4 different components $\mathbf{E} = (\mathbf{E^A}, \mathbf{E^N}, \mathbf{E^S}, \mathbf{E^C})$ as follows:

$\mathbf{E^A}$—ammonia (NH_3);
$\mathbf{E^S}$—sulphur dioxide (SO_2);
$\mathbf{E^N}$—nitrogen oxides ($NO + NO_2$);
$\mathbf{E^C}$—anthropogenic hydrocarbons.

The output of the model is mean monthly concentration of the following 3 pollutants:

s_1 —ozone (O_3);
s_2 —ammonia (NH_3);
s_3 —ammonium sulphate and ammonium nitrate ($NH_4SO_4 + NH_4NO_3$).

Table 1 Relative error and computational time for the approximate evaluation of $f_0 \approx 0.048$. The bold values denote the best relative error, i.e. the most accurate values

N	MSS-2S Rel. error	L-PROD Rel. error	L-FIB Rel. error	L-BIJ Rel. error
2^8	**1e−05**	2e−03	8e−04	2e−02
2^{10}	**4e−06**	2e−04	2e−04	8e−04
2^{14}	**2e−07**	1e−05	2e−05	3e−06
2^{16}	**2e−08**	4e−06	9e−06	4e−07

Table 2 Relative error for the approximate evaluation of the total variance $\mathbf{D} \approx 0.0002$. The bold values denote the best relative error, i.e. the most accurate values

N	MSS-2S Rel. error	L-PROD Rel. error	L-FIB Rel. error	L-BIJ Rel. error
2^8	**9e−04**	2e−02	3e−01	3e−01
2^{10}	**3e−04**	9e−02	2e−01	2e−02
2^{14}	**2e−05**	5e−03	3e−03	1e−03
2^{16}	**2e−06**	6e−04	3e−04	2e−03

Table 3 Relative error for estimation of sensitivity indices of input parameters using various stochastic approaches ($N \approx 65536$). The bold values denote the best relative error, i.e. the most accurate values

EQ	RV	MSS-2-S	L-PROD	L-FIB	L-BIJ
S_1	9e−01	5e−04	**1e−04**	4e−04	7e−04
S_2	2e−04	7e−02	1e−01	2e−01	**3e−02**
S_3	1e−01	1e−02	**2e−03**	3e−03	4e−03
S_4	4e−05	6e−01	**4e−03**	5e−01	2e−02
S_1^{tot}	9e−01	1e−03	**1e−04**	5e−04	5e−04
S_2^{tot}	2e−04	**3e−03**	4e−01	3e−01	2e−01
S_3^{tot}	1e−01	4e−03	**2e−03**	**2e−03**	6e−03
S_4^{tot}	5e−05	**1e−01**	7e−01	5e−01	2e−01

In our particular case we are interested in sensitivity studies of the mean monthly concentrations of ammonia in Milan. The domain under consideration is the 4-dimensional hypercubic domain $[0.5, 1]^4$).

The estimated quality is denoted by EQ. The results for relative errors for the approximate evaluation of the quantities f_0, total variance and first-order and total sensitivity indices using various stochastic approaches for numerical integration are presented in Tables 1, 2, and 3, respectively. The quantity f_0 is presented by 4-dimensional integral whereas the rest of quantities are presented by double dimensional integrals.

For the model function f_0 follows directly from Tables 1 that the best algorithm is MSS-2-S, followed by the L-BIJ and L-PROD. For the total variance D form Tables 2 for the maximum number of samples, follows that the best algorithm is MCA-MSS-2-S, followed by the L-PROD.

From the Table 3 it can be seen that for most sensitivity indices S_1, S_3, S_4, S_1^{tot} and S_3^{tot} the best relative error is produced by the proposed lattice rule with prime number of points and with product weights. The modified Sobol sequence MCA-MSS-2-S produce the best results for S_2^{tot} and S_4^{tot}, L-BIJ gives the best relative error for S_2 and L-FIB gives the best relative error for S_3^{tot}. Small in values sensitivity indices like S_2, S_4, S_2^{tot} and S_4^{tot} are very important for reliability of the model results. So there is no clearer winner in this case. However, for most of the cases Fibonacci based lattice rule and Bijective lattice rule give very close results and are outperformed by the new proposed lattice rule.

3.2 Sensitivity Studies with Respect to Chemical Reactions Rates

In this part we will study the sensitivity of the ozone concentration in Genova according to the rate variation of some chemical reactions: ## $1, 3, 7, 22$ (time-dependent) and $27, 28$ (time independent) reactions of the condensed CBM-IV scheme [17]. The simplified chemical equations of these reactions are as follows:

$$
\begin{aligned}
&[\#1]\ \ NO_2 + h\nu \Longrightarrow NO + O;\\
&[\#3]\ \ O_3 + NO \Longrightarrow NO_2;\\
&[\#7]\ \ NO_2 + O_3 \Longrightarrow NO_3;\\
&[\#22]\ \ HO_2 + NO \Longrightarrow OH + NO_2;\\
&[\#27]\ \ HO_2 + HO_2 \Longrightarrow H_2O_2;\\
&[\#28]\ \ OH + CO \Longrightarrow HO_2.
\end{aligned}
$$

The domain under consideration is the 6-dimensional hypercubic domain $[0.6, 1.4]^6$).

The results for relative errors for the approximate evaluation of the quantities f_0, total variance and first-order and total sensitivity indices using various stochastic approaches for numerical integration are presented in Tables 4, 5 and 6, respectively. Firstly, we should specify that the quantity f_0 is presented by 6-dimensional integral whereas the rest of quantities under consideration are presented by double dimensional integrals.

For the model function f_0 follows directly from Table 4 for the maximum number of samples, that the best algorithm is MSS-2-S, followed by L-PROD algorithm. For the total variance D form Table 5 for the maximum number of samples, follows that the best algorithm is again MSS-2-S, followed by L-PROD. From the Table 6 it can

Table 4 Relative error and computational time for the approximate evaluation of $f_0 \approx 0.27$. The bold values denote the best relative error, i.e. the most accurate values

	MSS-2-S	L-PROD	L-FIB	L-BIJ
N	Rel. error	Rel. error	Rel. error	Rel. error
2^6	**9e−05**	5e−03	2e−03	2e−01
2^{10}	**3e−06**	5e−03	1e−04	7e−03
2^{14}	**9e−08**	3e−04	4e−04	4e−05
2^{16}	**5e−07**	4e−05	3e−04	1e−05

Table 5 Relative error for the approximate evaluation of the total variance $\mathbf{D} \approx 0.0025$. The bold values denote the best relative error, i.e. the most accurate values

N	MSS-2-S Rel. error	L-PROD Rel. error	L-FIB Rel. error	L-BIJ Rel. error
2^6	**8e−03**	1e−01	4e+00	9e−01
2^{12}	**5e−04**	9e−02	5e−01	9e−02
2^{14}	**6e−06**	1e−02	1e−01	8e−04
2^{16}	**1e−04**	8e−04	2e−03	9e−04

Table 6 Relative error for estimation of sensitivity indices of input parameters using various stochastic approaches ($N \approx 65536$). The bold values denote the best relative error, i.e. the most accurate values

EQ	RV	MSS-2-S	L-PROD	L-FIB	L-BIJ
S_1	4e−01	2e−02	**3e−03**	4e−02	1e−02
S_2	3e−01	6e−02	**1e−03**	1e−02	2e−02
S_3	5e−02	8e−02	**3e−02**	5e−01	8e−02
S_4	3e−01	4e−03	**6e−03**	1e−02	7e−03
S_5	4e−07	2e+02	**2e+00**	3e+03	3e+03
S_6	2e−02	4e−02	**1e−02**	1e+00	**1e−02**
S_1^{tot}	4e−01	5e−02	**8e−03**	8e−02	1e−02
S_2^{tot}	3e−01	3e−02	**6e−03**	3e−02	2e−02
S_3^{tot}	5e−02	4e−02	**3e−02**	1e+00	5e−02
S_4^{tot}	3e−01	4e−02	**8e−03**	4e−01	2e−02
S_5^{tot}	2e−04	1e+00	**1e−01**	9e+01	9e+01
S_6^{tot}	2e−02	4e−02	**4e−03**	2e+00	8e−02
S_{12}	6e−03	7e−01	3e−01	3e+00	**2e−01**
S_{14}	5e−03	1e+00	**1e−02**	8e+00	1e+00
S_{24}	3e−03	1e+00	**4e−02**	1e+01	6e−01
S_{45}	1e−05	4e+00	**3e+00**	4e+01	4e+01

be seen that for almost all of the sensitivity indices, namely S_1, S_2, S_3, S_4, S_5, S_6, S_1^{tot}, S_2^{tot}, S_3^{tot}, S_4^{tot}, S_5^{tot}, S_6^{tot}, S_{14}, S_{24} and S_{45} the best relative error is produced by the proposed lattice rule with prime number of points and with product weights L-PROD. Only for S_{12} L-BIJ gives slightly better results than the proposed L-PROD and for S_6 L-FIB gives exactly the same relative error as L-PROD. Small in values sensitivity indices like S_5 and S_{45} are very important for reliability of the model results and this case we have a clear winner—the proposed algorithm. It is worth mentioning that while for S_5 the developed algorithm gives relative error 2 order better than MSS-2-S, in the case of S_{45} the two methods L-PROD and MSS-2-S produce relative errors of the same magnitude. It is important to mention that in the case of sensitivity analysis with respect to chemical reaction rates, where higher dimensional integrals appeared, the described algorithm outperforms the modified Sobol sequence.

To summarize, the benefits of the propoese mehod are the folowing:

- In the case of sensitivity study with respect to emission levels the propoesed method gives comparable results with the best modified Sobol sequence and

significantly improves the results by the Fibonacci lattice rule and Bijective lattice rule.
- The proposed method significantly improves the results by the modified Sobol sequence and other lattice methods especially in case of sensitivity study with respect to chemical reaction rates.
- The developed method gives the best results for smallest in value sensitivity indices, which are the most important for the reliability of the model results, with increasing the dimensionality of the integrals.
- The other benefit is that the computational complexity of the presented algorithm is linear, and it gives a very low relative error in less than a minute on a laptop.

4 Conclusion

In this paper a special lattice rule is constructed based on the rank-1 lattice rules with prime number of points and with product weights. It essentially improves the results by the Fibonacci based lattice rule and Bijective lattice rule. The obtained lattice rule with our optimal generating vector has an optimal rate of convergence for the corresponding class of functions which define the sensitivity indices in the multidimensional air pollution model. The proposed lattice rule improves the results by the other lattice rules and gives sufficient accuracy for most of the sensitivity indices and outperforms the results produced by the modified Sobol sequence with increasing the dimensionality of the quantities. The developed method gives the best results for smallest in value sensitivity indices with increasing the dimensionality of the integrals. The other benefit is that the computational complexity of the presented algorithm is linear, and it gives a very low relative error in less than a minute on a laptop. In the future work more efficient methods based on polynomial lattice rule will be developed. Although in the future work some limitations of this algorithm should be addressed connected with the more efficient choice of the generating vector. The obtained results will be important for model improvement, increase the reliability of results and identify the parameters and mechanisms that need to be examined more precisely.

Acknowledgments Venelin Todorov is supported by the Bulgarian National Science Fund under Project KP-06-N52/5 “Efficient methods for modeling, optimization and decision making” and by the Project KP-06-N62/6 “Machine learning through physics-informed neural networks”. The work is also supported by the Bulgarian National Science Fund under Young Scientists Project KP-06-M32/2 - 17.12.2019 “Advanced Stochastic and Deterministic Approaches for Large-Scale Problems of Computational Mathematics” and by the Project KP-06-Russia/17 “New Highly Efficient Stochastic Simulation Methods and Applications”, funded by the National Science Fund—Bulgaria.

References

1. Cools, R., Kuo, F., Nuyens, D.: Constructing embedded lattice rules for multivariate integration. SIAM J. Sci. Comput. **28**(6), 2162–2188 (2006). https://doi.org/10.1137/06065074X
2. Dimov, I.: Monte Carlo Methods for Applied Scientists, 291pp. World Scientific, New Jersey (2008)
3. Dimov, I.T., Georgieva, R.: Monte Carlo method for numerical integration based on sobol' sequences. In: Dimov, I., Dimova, S., Kolkovska, N. (eds.) Numerical Methods and Applications. Lecture Notes in Computer Science, vol. 6046, pp. 50–59. Springer, Berlin (2011)
4. Dimov, I.T., Georgieva, R.: Multidimensional sensitivity analysis of large-scale mathematical models. In: Iliev, O.P., et al. (eds.) Numerical Solution of Partial Differential Equations: Theory, Algorithms, and Their Applications. Springer Proceedings in Mathematics & Statistics, vol. 45, pp. 137–156. Springer, New York (2013)
5. Dimov, I.T., Georgieva, R., Ostromsky, T.Z., Zlatev, Z.: Advanced algorithms for multidimensional sensitivity studies of large-scale air pollution models based on sobol sequences. Comput. Math. Appl. **65**(3), 338–351 (2013). "Efficient Numerical Methods for Scientific Applications". Elsevier
6. Ferretti, F., Saltelli A., Tarantola, S.: Trends in sensitivity analysis practice in the last decade. Sci. Total Environ. **568**, 666–670 (2016). Special issue on Human and Biota Exposure, Elsevier
7. Gery, M., Whitten, G., Killus, J., Dodge, M.: A photochemical kinetics mechanism for urban and regional scale computer modelling. J. Geophys Res. **94**(D10), 12925–12956 (1989)
8. Haber, S.: Parameters for integrating periodic functions of several variables. Math. Comput. **41**(163), 115–129 (1983)
9. Homma, T., Saltelli, A.: Importance measures in global sensitivity analysis of nonlinear models. Reliab. Eng. Syst. Safety **52**, 1–17 (1996)
10. Kuo, F.Y., Nuyens, D.: Application of quasi-Monte Carlo methods to elliptic PDEs with random diffusion coefficients - a survey of analysis and implementation. Found. Comput. Math. **16**(6), 1631–1696 (2016)
11. Saltelli, A., Tarantola, S., Campolongo, F., Ratto,M.: Sensitivity Analysis in Practice: A Guide to Assessing Scientific Models. Halsted Press, New York (2004)
12. Sobol, I.: Numerical Methods Monte Carlo. Nauka, Moscow (1973)
13. Sobol, I.M.: Sensitivity estimates for nonlinear mathematical models. Math. Model. Comput. Exp. **1**(4), 407–414 (1993)
14. Veleva, E., Georgiev, I.R., Zheleva, I., Filipova, M.: Markov chains modelling of particulate matter (PM10) air contamination in the city of Ruse, Bulgaria. In: AIP Conference Proceedings, vol. 2302, no. 1, p. 060018. AIP Publishing LLC (2020)
15. Wang, Y., Hickernell, F.J.: An historical overview of lattice point sets. In: Monte Carlo and Quasi-Monte Carlo Methods 2000, Proceedings of a Conference held at Hong Kong Baptist University (2000)
16. Zaharieva, S.L., Georgiev, I.R., Mutkov, V.A., Neikov, Y.B.: Arima approach for forecasting temperature in a residential premises part 2. In: 2021 20th International Symposium INFOTEH-JAHORINA (INFOTEH), pp. 1–5. IEEE, Piscataway (2021)
17. Zlatev, Z.: Computer Treatment of Large Air Pollution Models. KLUWER Academic Publishers, Dorsrecht (1995)
18. Zlatev, Z., Dimov, I., Georgiev, K.: Modeling the long-range transport of air pollutants. IEEE Comput. Sci. Eng. **1**(3), 45–52 (1994)
19. Zlatev, Z., Dimov, I.T., Georgiev, K.: Three-dimensional version of the Danish Eulerian model. Z. Angew. Math. Mech. **76**(S4), 473–476 (1996)
20. Zlatev, Z., Dimov, I.T.: Computational and Numerical Challenges in Environmental Modelling. Elsevier, Amsterdam (2006)
21. Zlatev, Z., Dimov, I.T.: Using a digital twin to study the influence of climatic changes on high ozone levels in Bulgaria and Europe. Atmosphere **13**, 932 (2022). https://doi.org/10.3390/atmos13060932

On Pitfalls in Statistical Analysis for Risk Assessment of COVID-19

Tomomi Yamada, Hiroyuki Mori, Todd Saunders, and Tsuyoshi Nakamura

Abstract We describe pitfalls in statistical methods employed in epidemiological literature dealing with risk factors for COVID-19. Examples from the current literature are used to illustrate the pitfalls. Proper statistical techniques that illustrate correct statistical methods are then explained.

Keywords COVID-19 · Statistical methods · Pitfalls · Risk · Bias

1 Introduction

The objective of this study is to point out incorrect statistical analysis of data and interpretations of results we encountered when we reviewed epidemiological studies dealing with COVID-19.

Friedrich Nietzsche wrote, *"There are no facts, only interpretations"*[8]. And in truth, data is just a quantified fact of information. The data itself does not show the truth. The reader's interpretation will give the data a particular meaning, but it does not necessarily have to be true. Biased data produces a biased interpretation, however; there is another issue, incorrect analysis of unbiased data can lead to incorrect interpretation. To avoid incorrect analysis and interpretation, Statistics

T. Yamada
Department of Medical Innovation, Osaka University Hospital, Osaka, Japan
e-mail: tomomi.yamada@dmi.med.osaka-u.ac.jp

H. Mori
Department of Life and Creative Sciences, Nagasaki Women's College, Nagasaki, Japan
e-mail: mori@nagasaki-joshi.ac.jp

T. Saunders
Graduate School of Biomedical Science, Nagasaki University, Nagasaki, Japan

T. Nakamura (✉)
Faculty of Environmental Science, Nagasaki University, Nagasaki, Japan
e-mail: naka@nagasaki-u.ac.jp

© The Author(s), under exclusive license to Springer Nature Switzerland AG 2023
C. P. Kitsos et al. (eds.), *Statistical Modelling and Risk Analysis*,
Springer Proceedings in Mathematics & Statistics 430,
https://doi.org/10.1007/978-3-031-39864-3_17

has been developed. Statistics is a mathematical science for learning from data, and of measuring, controlling and communicating uncertainty [1]. Based on the statistics, epidemiologists help with study design, collection, and statistical analysis of data, as well as amend interpretation and dissemination of results. Epidemiology is a cornerstone of public health, and shapes policy decisions and evidence-based practice by identifying risk factors for disease and targets for preventive healthcare.

Because of the value we place in good statistical analysis, we were disappointed to find self-styled, incorrect statistical analysis published in leading journals such as *Nature* and *Lancet*. This implies that not only do authors lack exact knowledge of statistics; the reviewers do as well. We are afraid that readers of these journals will use those statistical methods, which have been accepted as correct to produce more incorrect analysis.

Here, we discuss four papers [6, 12, 17, 21]. Three of them deal with population-based cohort data obtained from linking several large databases, and one uses self-reported data of over two million participants. They all treat risk of COVID-19.

2 Methods

We classified incorrect methods used in the four studies into the following five categories: Cox and logistic models, Analysis of combined samples, Selection biases, Adjusting for confounders, and Biases due to measurement errors.

3 Results and Discussion

3.1 Use of Cox and Logistic Models

3.1.1 At Risk

Drefahl et al. [6] studied Cov-19 risk in 7.8 million individuals aged 20 or over in Sweden. The endpoint is deaths from COVID-19. 3126 endpoints were observed in their follow-up period March 13 – May 7, 2020. They used the Cox proportional hazards model: $\lambda(t|Z) = \lambda_0(t)exp(\beta^T Z)$ [4] with t being biological age in months, or more explicitly, $\lambda(age|Z) = \lambda_0(age)exp(\beta^T Z)$. Table 1 shows the results. The 1st column shows the name of variables, the 2nd the number of levels, the 3rd the range of the hazards for the levels by univariate Cox model analysis, and the 4th that by multivariate Cox model analysis. Age is not in Table 1, since age is not a covariate.

It is conspicuous that hazards are very small compared to other studies. Closely investigating their methods of data manipulation and statistical analysis, we conclude that the major cause of the small hazard estimates is the Cox model they

Table 1 Range of hazards for each variable (cited from Supplementary Table 6)

Variable	Levels	Univ.cox	Multiv.cox
Sex	2	1–1.7	1–2.1
Civil status	4	1–1.6	1–1.7
Education	3	1–1.2	1–1.3
Income	3	1–1.3	1–1.6
Birth country	4	1–5.1	1–2.8
Residence	2	1–4.5	1–4.6

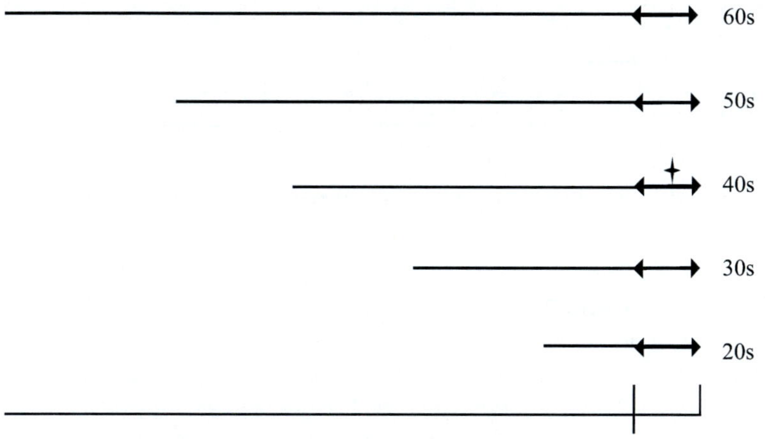

Fig. 1 Subjects are grouped into 10-year-old increments. Follow-up period is marked by arrows. Plus sign in 40s indicates an event

used, that is $\lambda(t|Z) = \lambda_0(t)exp(\beta^T Z)$ with t being biological age. The reason is explained using a simple example.

Figure 1 shows a calendar with the follow-up period marked by arrows, where subjects are grouped into 10-year-old increments. If a 45-year-old person reports a positive test, then those who are the same age as or older than the case, shown by vertical line in Fig. 2, comprise *at risk*. In survival analysis, or time-to-event data analysis, a hazard is calculated at each event time. In non-parametric tests such as the logrank test, the hazard is obtained as the reciprocal of the number of subjects at risk. In the Cox model, each subject is assigned a hazard $\lambda(t|Z) = \lambda_0(t)exp(\beta^T Z)$ consisting of covariates and parameters. Then, at each event time, say t, the hazard for the subject who experienced the event is divided by the sum of the hazards of the subjects being followed at t, or *at risk* at t.

However, those in their 50s and 60s on the vertical line are actually not *at risk* of COVID-19 death since they are exposed to COVID-19 10 or 20 more years later. In other words, their model regards those with no risk as *at risk*. Since the hazard is calculated as the reciprocal of the number of subjects at risk, the analysis resulted in small hazard estimates. Proper analysis is to define t as the days elapsed since March 13th and use age as a covariate.

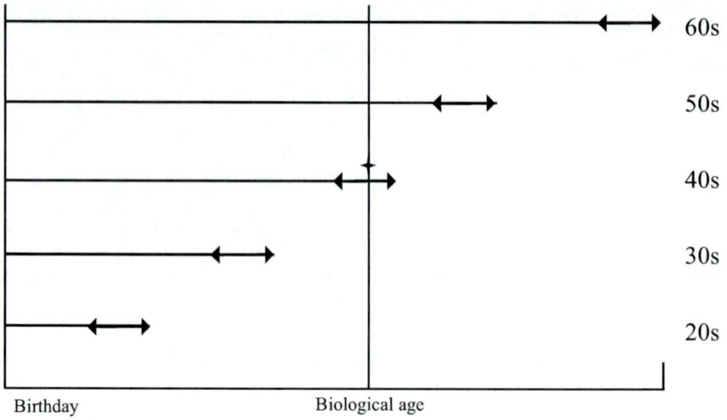

Fig. 2 Time scale is biological age. Follow-up period is marked by arrows. Plus sign denotes an event

3.1.2 Censored Sample

Drefahl et al. [6] stated *"Cox regression may even produce bias if there is substantive unaccounted spatiotemporal variation in the spread of the disease."* without any literature or evidence to support the idea. This is simply the authors' misconception. The assumption of the proportional hazards model is basically only *the proportionality of the hazards* that is usually confirmed using log-log survival, or log cumulative hazards, plot. *"Spatiotemporal variation in the spread of the disease"* might be concerned with the interpretation of estimated hazards, but is not concerned with the proportionality of hazards.

They further stated *"To assess this possibility, we estimated logistic regression models, which ignore the timing of the event and thus may rule out the possibility that the findings are driven by such patterns."* This is the authors' misunderstanding of the logistic model. Cox model is suitable for the censored survival data dealt with in their study, but applying a logistic model to a censored survival data is totally invalid [5]. In other words, the logistic model is applicable only to data without any censored cases.

Finally, they concluded *"We found no substantive difference between the Cox and the logistic regression."* This seems their subjective decision. What are the criteria for comparing odds and hazards and determining that they are not much different?

If *"substantive unaccounted spatiotemporal variation in the spread of the disease"* means low accuracy of the event times, that will result in underestimation of hazards. If the event times are accurate, the log-log plot or the log cumulative hazard plot is a simple method for confirming the proportionality of hazards.

3.2 Analysis of Combined Samples

Nguyen et al. [17] assessed the risk of COVID-19 among front-line health-care workers (HCW). Subjects are smartphone users who sent their demography, chronic disease, answers to daily questions regarding health status, and COVID-19 test results. Outcome is a report of a positive COVID-19 test. Their major finding is that the adjusted hazard ratio (HR) of reporting a positive COVID-19 test for HCW was 11.6. They performed the study for UK and USA separately, then simply combined the samples from two countries. Table 2 shows the results.

Briefly, 11.6 is a weighted average of 12.5 and 2.9. It is understood that 12.5 and 2.9 are, respectively, the hazard ratio of HCW compared to GM for UK and USA. But what does 11.6 mean exactly? Before combining them, it should be addressed why the risk for the USA is much lower than UK. Secondly, the combined risk should be affected by differences in living environment, particularly by the number of smartphone users in each country, the meaning of 11.6 is not straightforward. A proper method for combining samples is demonstrated by Nakamura et al. [16], where the endpoint of their study is *"severe case"* defined as either died or hospitalized in the ICU. The proportion of severe cases is termed *"severe rate"*. Figure 3 shows the number of COVID-19 cases (A) and severe cases (B) by week since February 20th, 2020. The first wave ended at week 12, but began to spread again at week 18. Weeks 0–15 will be denoted by Stage 0 and weeks 16–28 by Stage 1. To investigate relationships between the severe rates in Stage 0 and 1 (Table 3), a simple linear model $y = \alpha + \beta x + \varepsilon, \varepsilon \sim N(0, \sigma^2)$ was applied, where α is a constant, β is a regression coefficient, and x and y take values log (Severe rate) in Stage 0 and Stage 1, respectively.

Figure 4 shows a scatter plot between log(Severe rate) in Stage 0 and Stage 1. We have $\hat{\beta} = 1.7 (p < 0.001)$, $\hat{\alpha} = 0.18 (P = 0.59)$, and $R^2 = 0.94$. The results suggest it approximately holds that $y = 1.7x + \varepsilon$. Briefly, Severe rates in Stage 1 are approximately a function of those in Stage 0, irrespective of the levels of the variables. If we find a statistical model to represent the relationship, we may combine the cases of the two stages for more efficient analysis.

First, we define a binary variable $Stage$ as $Stage = 0$ for cases in Stage 0 and $Stage = 1$ for those in Stage 1. A stepwise logistic model using all variables

Table 2 HR for HCW compared to GM (general community), adjusted for sex, comorbidities, smoking, and BMI. (cited from Supplementary Table 6 [17])

Country	Occupation	Number of events	Person-days (in 1000)	Hazard ratio
UK	GM	3450	318,400	1
	HCW	1851	1309	12.5
USA	GM	173	1141	1
	HCW	71	145	2.9
UK+USA	GM	3623	319,541	1
	HCW	1922	1454	11.6

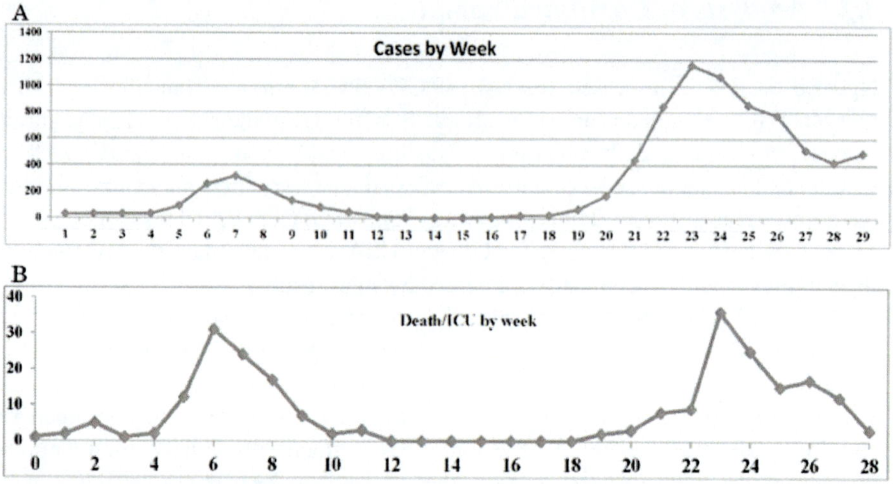

Fig. 3 Number of COVID-19 cases (**a**) and number of severe cases (**b**) by week from February 20th to September 10th, 2020

Table 3 Frequency of cases and severe rate by stage (cited from Table 3 [16])

		Stage 0		Stage 1	
Variable	Code: level	Freq.	Rate(%)	Freq.	Rate(%)
Age	20:0–29	351	0	3460	0
	40:30–49	511	1.57	2252	0.04
	60:50–69	364	8.24	1445	2.15
	80:70–99	268	29.1	1039	10.2
Sex	0:Female	674	6.08	3555	1.18
	1:Male	820	9.15	4639	2.07
Family status	0:With	879	5.8	5228	1.07
	1:No	249	7.63	2427	2.39
	2:Unclear	366	12.6	541	4.44
Workplace	0:Hosp./Sch.	240	0.83	1538	0.07
	1:Service	235	4.6	2207	0.41
	2:Indoor office	343	4.4	1838	0.38
	3:Unemployed	210	14.3	1520	5.66
	4:Unclear	466	12.5	1093	3.2
Comorbidity	0:No	1329	6.3	7387	1.1
	1:With	165	20	809	7.05

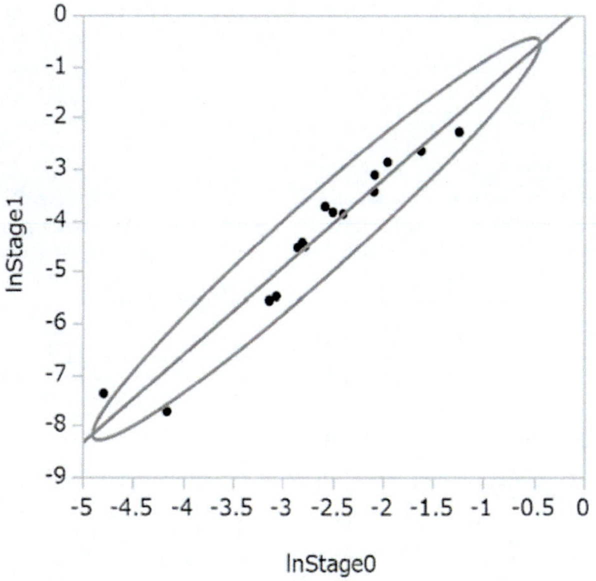

Fig. 4 Scatter plot between log (Severe rate) in Stage 0 and Stage 1 with 95% density ellipse

plus *Stage* as covariates is used to determine a best predictive model. The results indicate all variables are significant with $Stage = -1.24$. We also applied a logistic model with the variables excluding *Stage*. Distribution of the estimated probability of severe risk for the severe cases are shown in Fig. 5a, b, with *Stage* or without *Stage*, respectively. When *Stage* is excluded, the estimated risks for Stage 0 are similarly distributed as those for Stage 1 due to the same equation used. This is misleading, since the severe risk for Stage 0 is nearly five times higher than that for Stage 1. On the other hand, the distribution of the estimated risks for Stage 0 is closer to 1 than that for Stage 1 due to the inclusion of $Stage = -1.24$, or by taking into account the difference at the baseline risk between the Stages.

Concluding remark: It is necessary to confirm the possibility of combining samples and consider an appropriate method for combining samples. Simply combining samples might result in Simpson's paradox.

3.3 Adjusting for Confounder

Mutambudi et al. [12] investigated severe COVID-19 risk, defined as death with COVID-19 or testing-positive while an inpatient or in ICU, by occupational group. Their sample included 120,075 working participants aged 49 to 64 years in 2020, after excluding participants who died before 16 March 2020 ($n = 2067$) and those with missing data. Of these, the total 29.3% ($n = 35,127$) were classified as

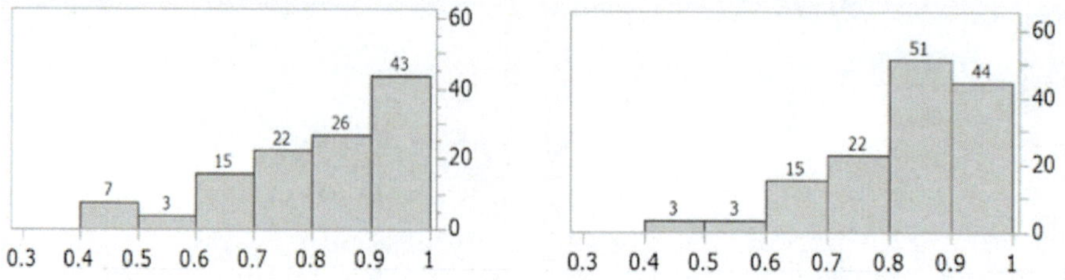

(a) Stage Excluded. Left: Stage=1; Right: Stage=0

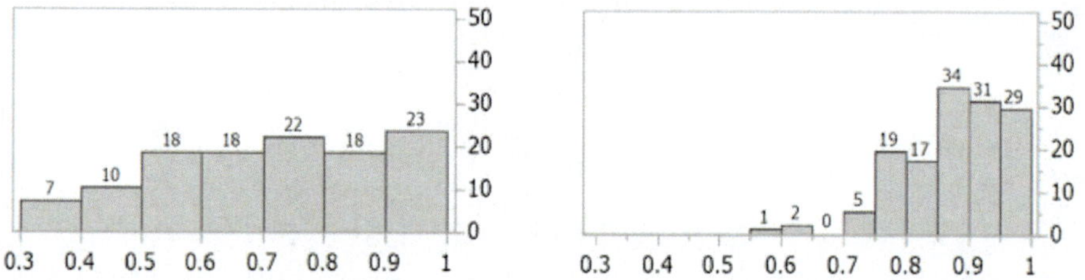

(b) Stage Included. Left: Stage=1; Right: Stage=0

Fig. 5 Distribution of estimated probability for severe cases. (**a**) Stage excluded. Left=1; Right: stage=0. (**b**) Stage included. Left=1; Right: stage=0

essential workers. The major finding is that essential workers have a higher risk of severe COVID-19. They used a Poisson regression model to obtain the occupation-specific risk adjusted for covariates such as demographic factors (age, sex, ethnicity, birth country), Socioeconomic deprivation (SEP) level, and education level. While demographic factors are confounders that should be adjusted for in estimating the risk for an occupation, SEP is a so-called *"intermediate factor"* in epidemiology that should not be adjusted for. Occupations affect SEP and, in turn, SEP affects the risk of severe COVID-19. Risk is underestimated by being adjusted for intermediate factors.

3.4 Correction for Selection Bias

Subjects of Nguyen et al. [17] are smartphone users who answer daily questions regarding health status and COVID-19 test results, as well as provide their private information. Since their outcome is *"a report of a positive COVID-19 test"* that requires receiving a test, they suspected a selection bias due to testing eligibility among subjects. To correct for the possible selection biases, the authors calculated the probability of receiving a COVID-19 test as a function of demographic factors, frequency of contact with possible COVID-19 patients, and symptoms at baseline. Unfortunately, since the endpoint, explanatory factors, and the equation used for calculating the probability are not reported, the performance of their prediction is unknown. Nevertheless, they performed Cox model with inverse probability weighted covariates.

If the chance of receiving the test is determined from the equation they obtained for each participant, and each participant receives the test randomly according to the probability, their attempt to correct for the selection biases might be successful. But given the complexity of human behavior, that assumption is unlikely. The selection bias is never corrected without exact information on why they take or do not take the tests. In fact, taking health checks are individuals' habits. A major factor influencing participation in a community health screening program is an individual's habits or beliefs about undergoing health screening, which depends on the living environment and experience (Kashiwazaki et al. [7], Okajima et al. [18]). If the selection bias could be corrected only by observable and measurable factors, then the randomization would not be necessary in clinical trials.

3.5 *Measurement Error*

For years, measurement error has been ignored because of the claim that its impact is *"not that bad"* [20]. We summarize the potentially severe consequences of measurement errors or misclassification, often overlooked in epidemiological analysis.

3.5.1 Biases Due to Measurement Error/Misclassification

Williamson et al. [21] studied the risk of COVID-19 using the primary care records of 17 million adults linked to 10,926 COVID-19-related deaths. The endpoint is COVID-19 related death between 2020/2/1-2020/5/6. The authors admit *"some COVID-19-related deaths may be misclassified, since non-confirmed cases are included."* However, they note that *"those errors should not have biased our hazard ratios, since this inaccuracy is likely to have reduced quickly as the number of deaths increased."* This is a naïve, false belief. The larger the number of cases, the closer the mean of them to the true value. However, they seem unaware of the important fact that the statistical power for detecting the difference also increases with increasing sample size. In effect, the bias in the results due to misclassifications in their statistical analysis remains irrespective of the sample sizes.

Consider a simple example to understand the fact: Let μ be a true value, and $X = \mu + \varepsilon$ be an observed value subject to error $\varepsilon \sim N(0, \sigma^2)$. If μ is observed n times, the mean of them $\bar{X}$ follows $N(\mu, \sigma^2/n)$. Therefore, as Williamson thought $\bar{X}$ approaches to μ as n increases. However, when $\bar{X}$ is used in statistical analysis, the sample size n is taken into account. Consider Z-test, $Z = \sqrt{n}(\bar{X} - \mu)/\sigma$ that follows $N(0, 1)$. In other words, the result of a statistical analysis using X is subject to biases due to the error, no matter how large n increases, unless an appropriate measure for correcting the bias (Nakamura [13], Carroll et al. [3]) is applied. This is why there has been "theory of measurement errors" for over 100 years.

Drefahl et al. [6] stated *"We cannot rule out some misclassification of COVID-19 deaths."* But they simply ignored the error in their analysis and interpretation of the results. Mutambudi et al. [12] stated that there exists occupational misclassification, since occupations are for 2006-10. They examined the agreement in occupation between baseline and follow-up using a sample. It turned out that agreement is between 45.8 and 76.1%. 45.8% is less than 50% by random allocation. The bias due to the substantial misclassification in occupation was simply ignored. Nguyen et al. [17] also ignored misclassification in the outcome *"a report of positive COVID-19 tests"*, discussed hereafter in 3.5.5. The four studies simply ignored the effects of biases due to measurement errors on the results. Nevertheless, they are published in leading journals, indicating that biases caused by measurement errors are not well understood.

3.5.2 Reliability Ratio

Consider a linear model $Y = \alpha + \beta Z + \varepsilon, \varepsilon \sim N(0, \sigma^2)$. Assume Z is not directly observed, instead $Z^* = Z + \varepsilon$ is observed, where $E(\varepsilon) = 0$. If we apply a linear model $Y = \alpha^* + \beta^* Z^* + \varepsilon^*$, ignoring the error, and obtain the ordinary estimate b^* for β^*, then it holds that $E(b^*) = \beta[var(Z)/var(Z^*)]$. Where, $var(Z)/var(Z^*) = Corr(Z^*, Z)^2$ is termed reliability ratio (RR) (Snedecor and Cochran [19]). For instance, if $RR = 0.7$, ignoring the measurement errors causes an attenuation by 30%. Nakamura [14] performs a simulation study to examine attenuation due to measurement errors in the Cox proportional hazards model to find that the attenuation with the Cox model is more serious than that with the linear model when RR is the same.

3.5.3 Adjusting for Unbalanced Confounder

Nakamura et al. [15] consider a proportional hazards model

$$\lambda(t|\Delta, Z) = \lambda_0(t)exp(\beta\Delta + \beta_z Z), \Delta = 0 \text{ or } 1$$

to estimate the treatment effect β adjusting for the confounding effect with Z. Z's for $\Delta = 0$ are sampled from a triangular distribution with support $(0, 12^{1/2})$ that has a density $2/12^{1/2}$ at $Z = 0$ and 0 at $Z = 2^{1/2}$. On the other hand, Z's for $\Delta = 1$ are sampled from a triangular distribution that has a density 0 at $Z = 0$ and $2/12^{1/2}$ at $Z = 12^{1/2}$. Figure 6 shows the triangular distributions. The sample size is 150 for each. $var(Z) = 1$ for the combined samples (Fig. 6). Z's are considerably unbalanced between the two groups. As before, we assume Z is subject to error and $Z^* = Z + \varepsilon, \varepsilon \sim N(0, \sigma^2)$, is observed. Simply replacing Z with Z^* in the

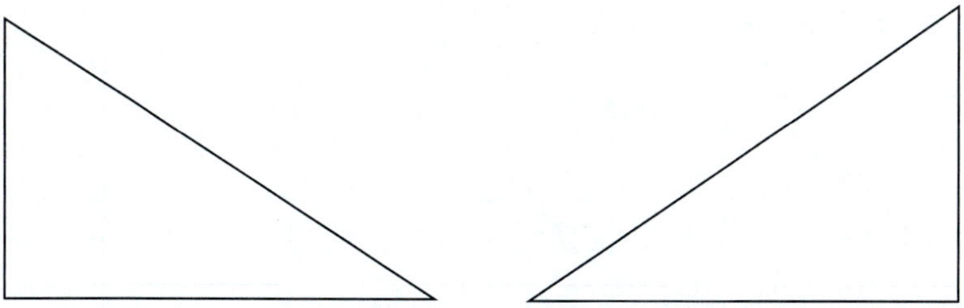

Fig. 6 Two triangle distributions with support $(0, 12^{1/2})$ for $\Delta = 0$ and 1

Table 4 β^* and β_z^* are average of estimates of β and β_z obtained from repetitions. (cited from Table 5 [15].)

σ	RR	β^*	β_z^*
0.4	0.86	−0.10	0.71
0.5	0.80	−0.03	0.64
0.6	0.74	0.05	0.56
0.7	0.67	0.11	0.49
0.8	0.61	0.19	0.43

estimation of β results in a biased estimate. This is examined by a simulation with $\beta = -0.3$, $\beta_z = 1$, and $\sigma = 0.4 \sim 0.8$. The results are presented in Table 4.

β^* and β_z^* are the average of 100 estimates of β and β_z, respectively, obtained from 100 repetitions. It is striking that the attenuated estimate β^* indicates almost no effect of the treatment for $\sigma = 0.5$ or 0.6, and even reverse effect of the treatment is observed for larger σ, or $RR < 0.7$. In conclusion, $RR > 0.7$ is necessary for adjustment for the unbalanced confounders to be effective.

3.5.4 Decreasing Power

Let $Q(Z, Y, n)$ denote a test with a covariate Z, an outcome Y and a sample of size n. We assume Z and Y are subject to measurement errors and $Z^* = Z + \varepsilon$ and $Y^* = Y + \delta$ are available. Z^* and Y^* are often called *surrogate* of Z and Y, respectively. It holds that *asymptotic relative efficiency* (ARE) of a test $Q(Z^*, Y, n)$ with respect to $Q(Z, Y, n)$ is

$$ARE(Z^*|Z) = Corr(Z^*, Z)^2$$

That is, the power of $Q(Z^*, Y, N)$ asymptotically equals that of $Q(Z, Y, n)$ when $N = n/Corr(Z^*, Z)^2$. If $Corr(Z^*, Z) = 0.7$, then $N = n/0.49$; twice a large sample size is required to attain the same statistical power when Z^* is used. The formula holds for the linear, logistic and Cox models (Lagakos [9]).

A similar formula holds for an outcome variable, too. Let Y be a binary outcome, $Q(Z, Y, n)$ be a test using a logistic model and Y^* be a surrogate subject to misclassifications. Then it holds (Yamada et al. [22]) that $ARE(Y^*|Y)$ of a test

$Q(Z, Y^*, n)$ with respect to $Q(Z, Y, n)$ is

$$ARE(Y^*|Y) = Corr(Y^*, Y)^2$$

The formula is extended to the case where both Z and Y are subject to measurement errors (Misumi et al. [11]) when Q is based on a logistic model:

$$ARE(Z^*, Y^*|Z, Y) = Corr(Z^*, Z)^2 Corr(Y^*, Y)^2$$

That is, when $N = n/\{Corr(Y^*, Y)^2 Corr(Z^*, Z)^2\}$, the power of $Q(Z^*, Y^*, N)$ asymptotically equal that of $Q(Z, Y, n)$.

3.5.5 Sensitivity and Specificity

Since the outcome *"a report of a positive COVID-19 test"* of Nguyen et al. [17] is conditional upon receiving a test, there could be information bias. To alleviate this issue, they use a logistic model, termed *symptom-based classifier*, for likelihood of COVID-19 infection described in Menni et al. [10]. Briefly, the prediction model uses such factors as age, sex, loss of smell/taste, persistent cough, fatigue and skipped meals. They obtained HR 2.05 for HCW using the outcome estimated by the prediction model. HR 2.05 is far lower compared to 11.6 obtained based on the reports of a positive COVID-19 test. They discuss, however, neither the reason nor the implication of the discrepancy. We suspect the misclassification in the prediction could be one of the reasons for the low HR. RR in their analysis is obtained as follows. They stated the sensitivity and specificity of their model is 0.65 and 0.78, respectively. Denoting the infected rate in population by p, we have Table 5. Further denoting the sensitivity and specificity by α and β, respectively, we have Table 6.

3.5.6 2 × 2 Misclassification Model

To obtain, $Corr(Z, X)$ in Table 6, consider a simplified notation in Table 7, where r, s, t, u are non-negative and $r + s + t + u = 1$.

Table 5 Sensitivity=0.65, specificity=0.78, and infection rate in population=p

		Predicted COVID-19		
		+	-	
Infected	+	$0.65p$	$(1-0.65)p$	p
COVID-19	-	$(1-0.78)(1-p)$	$0.78(1-p)$	$1-p$

Table 6 X=predicted, $Z = true$, 1=infected

		X		
		1	0	
Z	1	αp	$(1-\alpha)p$	p
	0	$(1-\beta)(1-p)$	$\beta(1-p)$	$1-p$

Table 7 Simplified Table 6

		X		
		1	0	
Z	1	r	s	p
	0	t	u	$1-p$

Lemma 1 $RR(X|Z) = Corr(Z, X)^2 = (ru - st)^2/\{(r+t)(s+u)(r+s)(t+u)\}$.

Proof *The result follows from* $var(X) = (r+t)(s+u)$, $var(Z) = (r+s)(t+u)$ *and* $Cov(Z, X) = ru - st$.

Theorem 1 $RR(X|Z) = \delta^2 p(1-p)/\{\theta(1-\theta)\}$, where $\delta = \alpha + \beta - 1$ and $\theta = \beta - \delta p$.

Proof Assigning r, s, t and u with the corresponding entity in Table 6, we have

$$\begin{aligned} RR(X|Z) &= \frac{\{\alpha p\beta(1-p) - (1-\alpha)p(1-\beta)(1-p)\}^2}{p(1-p)\{\alpha p + (1-\beta)(1-p)\}\{(1-\alpha)p + \beta(1-p)\}} \\ &= \frac{p(1-p)\{\alpha\beta - (1-\alpha)(1-\beta))\}^2}{\{1-\beta+(\alpha+\beta-1)p\}\{\beta+(1-\alpha-\beta)p\}} \\ &= \frac{p(1-p)(\alpha+\beta-1)^2}{\{1-\beta+(\alpha+\beta-1)p\}\{\beta-(\alpha+\beta-1)p\}} \\ &= \frac{p(1-p)\delta^2}{(\beta-\delta p)(1-\beta+\delta p)} \\ &= \frac{\delta^2 p(1-p)}{\theta(1-\theta)} \end{aligned}$$

In their study, positive and negative cases are 7,104 and 11,297, respectively, therefore $p = 0.386$. Assigning $\alpha = 0.65$, $\beta = 0.78$, we have $\delta = 0.65+0.78-1 = 0.43$, $\theta = 0.78-0.43\times0.386 = 0.614$. Therefore, $RR(X|Z) = \delta^2 p(1-p)/\{\theta(1-\theta)\} = \{0.43^2 \times 0.386(1-0.386)\}/\{0.614(1-0.614)\} = 0.185$. RR is quite small but that alone doesn't seem to explain the large discrepancy. Since the subjects of Menni et al. [10] are different from those of Nguyen et al. [17], the actual sensitivity and specificity are not known. They might be far lower due to the *transportability bias*. (Bareinboim et al. [2]).

4 Conclusion

Incorrect statistical methods used for epidemiological studies on the risk of COVID-19 in some leading journals are reported and proper statistical methods are described. We hope those journals will include experts in epidemiological statistics as reviewers so that epidemiological papers can show reliable and useful results for preventive measures.

Acknowledgments The authors are indebted to Yoshiaki Nose for insightful and encouraging comments from clinical medicine.

References

1. American Statistical Association. https://www.amstat.org/
2. Bareinboim, E., Pearla, J.: Causal inference and the data-fusion problem. Proc. Natl Acad. Sci. **113**(27), 7345–7352 (2016). www.pnas.org/cgi/doi/10.1073/pnas.1510507113
3. Carroll, R.J., Ruppert, D., Stefanski, L.A.: Measurement Error in Nonlinear Models. Chapman and Hall, London (1995)
4. Cox, D.R.: Regression models and life-tables. J. R. Stat. Soc. Ser. B **34**, 187–202 (1972)
5. Cox, D.R.: The analysis of multivariate binary data. J. R. Stat. Soc. Ser. C **21**(2), 113–120 (1972)
6. Drefahl, S., Wallace, M., Mussino, E.: A population-based cohort study of socio-demographic risk factors for COVID-19 deaths in Sweden. Nat. Commun. **11**, 5097 (2020). https://doi.org/10.1038/s41467-020-18926-3
7. Kashiwazaki, H., Moriyama, M., Sato, H., et al.: Factors effecting the use of mass health examination in an island population (in Japanese with English abstract). Japanese J. Publ. Health. **29**, 385–391 (1982)
8. Kaufmann, W.: The Portable Nietzsche. Penguin Classics, London (1977). ISBN-13:978-0140150629
9. Lagakos, S.W.: Effects of mismodelling and mismeasuring explanatory variables on tests of their association with a response variable. Stat. Med. **7**, 257–274 (1988)
10. Menni, C., Valdes, A.M., Maxim, B., et al.: Real-time tracking of self-reported symptoms to predict potential COVID-19. Nat. Med. **26**, 1037–1040 (2020). https://doi.org/10.1038/s41591-020-0916-2
11. Misumi, M., Yamada, T., Nakamura, T., Nose, Y.: Sample size determination in genetic disease association studies when the response variable is subject to misclassification and a surrogate covariate is used. In: Berhardt, L.V. (ed.) Advances in Medicine and Biology, chap. 5, vol. 5. Nova Science Publishers, Hauppauge (2010)
12. Mutambudzi, M., et al.: Occupation and risk of severe COVID-19: prospective cohort study of 120 075 UK Biobank participants. Occup. Environ. Med. **78**(5), 307–314 (2021)
13. Nakamura, T.: Corrected score function of errors-in-variables models: methodology and applications to generalized linear models. Biometrika **77**, 127–137 (1990)
14. Nakamura, T.: Proportional hazards model with covariates subject to measurement error. Biometrics **48**(3), 829–838 (1992). https://doi.org/10.2307/2532348
15. Nakamura, T., Akazawa, K.: Corrected likelihood for proportional hazards measurement error model and its application. Environ. Health Perspect. **102**(suppl 8), 21–24 (1994)
16. Nakamura, T., Mori, H., Saunders, T., Chishaki, H., Nose, Y.: Impact of workplace on the risk of severe COVID-19. Front. Public Health **9**, 731239 (2022). https://doi.org/10.3389/fpubh.2021.731239
17. Nguyen, L.H., Drew, D.A., Graham, M.S., et al.: Risk of COVID-19 among front-line health-care workers and the general community: a prospective cohort study. Lancet Public Health **5**, e475-83 (2020). https://doi.org/10.1016/S2468-2667(20)30164-X
18. Okajima, S., Mine, M., Nakamura, T.: Mortality of registered a-bomb survivors in Nagasaki, Japan, 1970–1984. Radiation Res. **103**, 419–431 (1985)
19. Snedecor, G.W., Cochran, W.G.: Statistical Methods. Iowa State University Press, Ames (1967)

20. Wallace, M.: Analysis in an imperfect world. Significance **17**, 14–19 (2020). https://rss.onlinelibrary.wiley.com/doi/epdf/10.1111/j.1740-9713.2020.01353.x
21. Williamson, E.J., Walker, A.J., Bhaskaran, K.: Factors associated with COVID-19-related death using OpenSAFELY. Nature **584**, 430–436 (2020). https://www.nature.com/articles/s41586-020-2521-4#Sec1
22. Yamada, T., Kinukawa, N., Nakamura, T., Nose, Y.: Simulation program for power and sample size determination in logistic analysis of single nucleotide polymorphisms when the response variable is subject to misclassification. Comput. Methods Programs Biomed. **96**, 42–48 (2009). https://doi.org/10.1016/j.cmpb.2009.03.007